Fenugreek

Medicinal and Aromatic Plants – Industrial Profiles

Individual volumes in this series provide both industry and academia with in-depth coverage of one major medicinal or aromatic plant of industrial importance.

Edited by Roland Hardman

Fenugreek

The genus *Trigonella*

Edited by
Georgios A. Petropoulos

CRC Press
Taylor & Francis Group
Boca Raton London New York

CRC Press is an imprint of the
Taylor & Francis Group, an **informa** business

A TAYLOR & FRANCIS BOOK

CRC Press
Taylor & Francis Group
6000 Broken Sound Parkway NW, Suite 300
Boca Raton, FL 33487-2742

First issued in paperback 2019

© 2002 by Taylor Francis Group, LLC
CRC Press is an imprint of Taylor & Francis Group, an Informa business

Typeset in 10/12 Garamond by
Newgen Imaging Systems (P) Ltd, Chennai, India

No claim to original U.S. Government works

ISBN-13: 978-0-415-29657-1 (hbk)
ISBN-13: 978-0-367-39590-2 (pbk)

British Library Cataloguing in Publication Data
A catalogue record for this book is available from the British Library

Library of Congress Cataloging in Publication Data
Fenugreek: the genus Trigonella / edited by George A. Petropoulos.
    p. cm – (Medicinal and aromatic plants – industrial profiles)
    ISBN 0-415-29657-9 (hbk.)
    1. Fenugreek. I. Petropoulos, George A. II. Series.

SB317 .F44 F45 2002
633.8′8—dc21   2002072359

Visit the Taylor & Francis Web site at
http://www.taylorandfrancis.com

and the CRC Press Web site at
http://www.crcpress.com

Contents

Figures

Tables

Contributors

Molham Al-Habori, Faculty of Medicine and Health Sciences, University of Sana'a, Sana'a, Republik of Yemen.

Christos V. Fotopoulos, National Agricultural Research Foundation (NAgReF), 4 Micropoulou str. 14121 N. Iraklio, Athens, Greece.

C.N. Giannopolitis, Benaki Phytopathological Institute, Weed Science Department, Greece.

Panagiotis Kouloumbis, National Agricultural Research Foundation (NAgReF), Athens Soil Science Institute, Greece.

George Manicas, 32, Analipseos str., 15235 Vrilissia, Greece.

Georgios A. Petropoulos, 4, Antiopis str., 173 43 Athens, Greece.

Amala Raman, King's College London, Department of Pharmacy, United Kingdom.

Helen Skaltsa, School of Pharmacy, Department of Pharmacognosy and Chemistry of Natural Compounds, University of Athens, Panepistimiopolis, Zografou, GR-15771, Athens, Greece.

Caroline G. Spyropoulos, University of Athens, Department of Biology, Institute of General Botany, Athens, Greece.

Preface to the series

There is increasing interest in industry, academia and the health sciences in medicinal and aromatic plants. In passing from plant production to the eventual product used by the public, many sciences are involved. This series brings together information which is currently scattered through an ever increasing number of journals. Each volume gives an in-depth look at one plant genus, about which an area specialist has assembled information ranging from the production of the plant to market trends and quality control.

Many industries are involved such as forestry, agriculture, chemical, food, flavour, beverage, pharmaceutical, cosmetic and fragrance. The plant raw materials are roots, rhizomes, bulbs, leaves, stems, barks, wood, flowers, fruits and seeds. These yield gums, resins, essential (volatile) oils, fixed oils, waxes, juices, extracts and spices for medicinal and aromatic purposes. All these commodities are traded worldwide. A dealer's market report for an item may say 'Drought in the country of origin has forced up prices'.

Natural products do not mean safe products and account of this has to be taken by the above industries, which are subject to regulation. For example, a number of plants which are approved for use in medicine must not be used in cosmetic products.

The assessment of safe to use starts with the harvested plant material which has to comply with an official monograph. This may require absence of, or prescribed limits of, radioactive material, heavy metals, aflatoxin, pesticide residue, as well as the required level of active principle. This analytical control is costly and tends to exclude small batches of plant material. Large scale contracted mechanized cultivation with designated seed or plantlets is now preferable.

Today, plant selection is not only for the yield of active principle, but for the plant's ability to overcome disease, climatic stress and the hazards caused by mankind. Such methods as *in vitro* fertilization, meristem cultures and somatic embryogenesis are used. The transfer of sections of DNA is giving rise to controversy in the case of some end-uses of the plant material.

Some suppliers of plant raw material are now able to certify that they are supplying organically-farmed medicinal plants, herbs and spices. The Economic Union directive (CVO/EU No 2092/91) details the specifications for the *obligatory* quality controls to be carried out at all stages of production and processing of organic products.

Fascinating plant folklore and ethnopharmacology leads to medicinal potential. Examples are the muscle relaxants based on the arrow poison, curare, from species of *Chondrodendron*, and the anti-malarials derived from species of *Cinchona* and *Artemisia*.The methods of detection of pharmacological activity have become increasingly reliable and specific, frequently involving enzymes in bioassays and avoiding the use of laboratory animals. By using bioassay linked fractionation of crude plant juices or extracts, compounds can be specifically targeted which, for example, inhibit blood platelet aggregation, or have anti-tumour, or anti-viral, or any other required activity. With the assistance of robotic devices, all the members of a genus may be readily screened. However, the plant material must be *fully* authenticated by a specialist.

The medicinal traditions of ancient civilizations such as those of China and India have a large armamentaria of plants in their pharmacopoeias which are used throughout South-East Asia. A similar situation exists in Africa and South America. Thus, a very high percentage of the World's population relies on medicinal and aromatic plants for their medicine. Western medicine is also responding. Already in Germany all medical practitioners have to pass an examination in phytotherapy before being allowed to practise. It is noticeable that throughout Europe and the USA, medical, pharmacy and health related schools are increasingly offering training in phytotherapy.

Multinational pharmaceutical companies have become less enamoured of the single compound magic bullet cure. The high costs of such ventures and the endless competition from 'me too' compounds from rival companies often discourage the attempt. Independent phytomedicine companies have been very strong in Germany. However, by the end of 1995, eleven (almost all) had been acquired by the multinational pharmaceutical firms, acknowledging the lay public's growing demand for phytomedicines in the Western World.

The business of dietary supplements in the Western World has expanded from the health store to the pharmacy. Alternative medicine includes plant-based, products. Appropriate measures to ensure the quality, safety and efficacy of these either already exist or are being answered by greater legislative control by such bodies as the Food and Drug Administration of the USA and the recently created European Agency for the Evaluation of Medicinal Products, based in London.

In the USA, the Dietary Supplement and Health Education Act of 1994 recognized the class of phytotherapeutic agents derived from medicinal and aromatic plants. Furthermore, under public pressure, the US Congress set up an Office of Alternative Medicine and this office in 1994 assisted the filing of several Investigational New Drug (IND) applications, required for clinical trials of some Chinese herbal preparations. The significance of these applications was that each Chinese preparation involved several plants and yet was handled as a *single* IND. A demonstration of the contribution to efficacy, of *each* ingredient of *each* plant, was not required. This was a major step forward towards more sensible regulations in regard to phytomedicines.

My thanks are due to the staffs of Harwood Academic Publishers and Taylor & Francis who have made this series possible and especially to the volume editors and their chapter contributors for the authoritative information.

Roland Hardman

Preface

In recent decades increasing attention has been paid in utilization and consumption of natural and traditional products (foods, flavours, colours, perfumes, phytotherapeutics etc.), because modern scientific knowledge and technologies have revealed that many chemical products of synthetic origin of this kind are responsible for a lot of new hazards and disorders for human beings.

The plant species of the genus *Trigonella* and especially that of *T. foenum-graecum* L. (fenugreek) is a good example, which has been used traditionally to cover such human needs. Fenugreek is cultivated all over the world and mainly in India and the Mediterranean countries as chemurgic, cash and good renovator of soil crop and as a multi-purpose legume, is used as forage, food, spice, perfume, insect repellent, dye, herbal medicine etc.

The biological and pharmaceutical actions of fenugreek are attributed to the variety of its constituents including steroids (diosgenin), alkaloids (trigonelline), flavonoids (luteolin), coumarins, aminoacids (hydroxyisoleucine), mucilage (galactomannan), volatile constituents (HDFM), fixed oils and other substances.

Species of the genus *Trigonella* and particularly fenugreek are well known for their pungent aromatic, high nutritive and multi-therapeutical properties and serve culinary, medicinal and industrial purposes.

As there is today an emerging change in food habits preference for natural colouring, flavouring and revolution in packaging, fenugreek could contribute to this direction, as its seeds are a component of many curry preparations and are used to colour and flavour food, stimulate appetite and help digestion.

Fenugreek is one of the oldest known medicinal plants from ancient times and even Hippocrates thought highly of it. Fenugreek seeds which are described in the Greek and Latin Pharmacopoeias are said to have anti-diabetic activity and hypocholesterolaemic effects and have been reported to possess a curative gastric anti-ulcer action and anti-fertility and anti-nociceptive effects. The therapeutic efficacy of fenugreek extracts in providing sedation has been proved by many pharmacological and clinical experiments. So, many of its actions as remedy have been confirmed and the mechanisms of their activity are being studied. Also, some other properties of fenugreek which have been reported but received less attention include anti-cancer, anti-bacterial, anthelmintic, anti-cholinergic, wound healing activities, etc.

Fenugreek seed as a source of diosgenin, that is the base for the production of the oral contraceptives and rich in protein and fixed oils, could make a two-fold economic contribution to the world's increasing population problems, by assisting in birth control and at the same time, providing additional food, especially for people, where meatless diets are customary for cultural and religious reasons.

Finally, it is doubtful if any other plant crop, while saving energy by fixation of atmospheric nitrogen, has such potential for making a major contribution to the world's food supply, to reduce hunger, improve health care and help population control.

Georgios A. Petropoulos

Acknowledgments

I would like to thank the numerous people who helped to make this work possible. In particular I appreciate Dr R. Hardman for his continuous advice and helpful suggestions, Dr Anthony Dweck, Research Director of Peter Black Toiletries and Cosmetics Ltd. for providing a data base of references on the use and history of fenugreek and Demetrios Cotarides for his assistance with the drawings.

Finally I am indebted to my family for their continuous encouragement.

1 Introduction

Georgios A. Petropoulos

This introductory chapter deals with a brief analysis of the history, world cultivated area, main uses, needs for research and future trends of the most important species of the genus *Trigonella* and especially that of *T. foenum-graecum* (fenugreek).

History

Plants of the genus *Trigonella* and particularly of the cultivated species *T. foenum-graecum* (fenugreek) were known and used for different purposes in ancient times, especially in Greece and Egypt (Rouk and Mangesha, 1963). In North Africa it has been cultivated around the Saharan oases since very early times (Duke, 1986).

Hidvegi *et al.* (1984) report that references to the utilization of fenugreek are found as far back as 1578, when detailed information on the plant is given in the famous Kolozsvar Herbarium compiled by Melius (1578). In this Transylvanian Herbarium the 'warming and very drying' nature of fenugreek and its antique sources are emphasized. Fenugreek seeds were found in the tomb of Tutankhamun (Manniche, 1989). Portius Cato, a Roman authority on animal husbandry in the second century BC ordered *foenum-graecum*, that was today's fenugreek, to be shown as fodder for oxen (Fazli and Hardman, 1968). Antiochus Epiphanes, King of Syria, and all those who entered the gymnasium to witness the games were anointed with perfumes from golden dishes that contained fenugreek and other aromatic plants (Leyel, 1987). Leaves of fenugreek were one of the components of the celebrated Egyptian Incense *Kuphi*, a holy smoke used in fumigation and embalming rites (Rosengarten, 1969). Miller (1969) reports that fenugreek was a spice plant mentioned in classical texts.

Historically, fenugreek is one of the oldest known medicinal plants and even Hippocrates thought highly of it (Lust, 1986; Schauenberg and Paris, 1990). There is a prescription for the rejuvenational properties of fenugreek of Pharaonic date (Manniche, 1989). Fenugreek was first introduced into Chinese medicine in the Sung dynasty, AD 1057 (Jones, 1989). Dioscorides, a greek physician of Anazarbus in Cilicia, father of Pharmacology, at AD 65, in his examination of the definition and function of spices in his Materia Medica, writes that fenugreek is an active compound of ointments (Miller, 1969). He also describes a concoction of fenugreek seeds to treat the vulva. In the seventeenth century fenugreek seeds were recommended to help expel the placenta of women after giving birth (Howard, 1987). The herb has long been a favourite of the Arabs and it was studied at the School of Salermo by Arab physicians (Stuart, 1986). Fenugreek was known and cultivated as forage in ancient Greece. Theophrastus had given it the greek names Βουκέρας (Voukeras) and Τῆλις (Tilis) and the oil produced from it was called *τῆλιον ἔλαιον* (oil of Tilis). Probably fenugreek is one of the forages known to have been cultivated before the era of recorded history. As a fodder plant, it is said to be the Hedysarum of

Theophrastus and Dioscorides (Leyel, 1987). Dioscorides also says that the Egyptians called it 'itasin' (Manniche, 1989). In the Middle Ages it is recorded that fenugreek was added to inferior hay, because of its peculiar pleasant smell (Howard, 1987).

Fenugreek was introduced into Central Europe at the start of the ninth century (Schauenberg and Paris, 1990), according to Fazli and Hardman (1968) Charlemagne encouraged its cultivation in this area. Rosengarten (1969) reports that the Romans obtained the plant from the Greeks, and that it became a commercial commodity of the Roman Empire (Miller, 1969), while Stuart (1986) and Howard (1987) support the contention that Benedictine monks introduced the plant into medieval Europe. However, it is not mentioned in any herbal literature until the sixteenth century, when it was recorded as grown in England.

Cultivated area

Furry (1950) describes five cultivated species of the genus *Trigonella* as: *T. foenum-graecum, T. caerulea, T. polycerata, T. monspeliaca* and *T. suavissima*, while in Flora European (Ivimey-Cook, 1968) only two species to be cultivated are reported: *T. foenum-graecum* and *T. caerulea*; the last one has also been reported as cultivated by Uphof (1968). However, statistics of the cultivated area for forage and seed production are not available, except for the *T. foenum-graecum* (fenugreek). Fenugreek has been reported as a cultivated crop in Portugal, Spain, United Kingdom, Germany, Austria, Switzerland, Greece, Turkey, Egypt, Sudan, Ethiopia, Kenya, Tanzania, Israel, Lebanon, Morocco, Tunisia, India, Pakistan, China, Japan, Russia, Argentine and the United States of America (Rouk and Mangesha, 1963; Fazli and Hardman, 1968; Rosengarten, 1969). At the present time fenugreek is an important cash crop in India (the leading fenugreek producing country), Morocco, China, Pakistan, Spain, Tunisia, Turkey, Lebanon, Israel, Egypt, Ethiopia, Kenya, Tanzania etc. (Smith, 1982; Edison, 1995).

As far as the world cultivated area of fenugreek and the annual production of seed are concerned, statistics are very limited and scattered, as the area seeded with fenugreek is relatively small and not recorded by the agricultural statistics of different countries. In spite of this, the following analysis based on the exported quantities of the principal producing countries, the domestic use of fenugreek and the existing statistics of the cultivated area for some countries, represents a reasonably accurate assessment of the world production and cultivated area of fenugreek.

So, taking into consideration that:

1   The cultivated area of fenugreek in India, an average for the last twenty years (1975–95), accounts for 34,534 ha with a production of 41,530 tons and an export of 4203 tons, that is domestic use accounts for 90 per cent of the production (Anonymous, 1996).

2   Recently, there has been an increase in the export of fenugreek from India: in 1994–95 it accounted for 7,956 tons (Anonymous, 1996). According to Edison (1995) India claims 70–80 per cent of the world export in fenugreek. This means that the world export of fenugreek until 1995 fluctuated around 10,500 tons, and export from the other countries mentioned above can be estimated as approaching 2,700 tons. According to the forementioned considerations, the cultivated area from these countries accounts for about 22,000 ha with a production of 26,700 tons.

3   These considerations permit us to estimate that in the world, the annually cultivated area of fenugreek amounts to roughly 57,000 ha with a seed production of 68,000 tons.

The wide distribution of fenugreek is indicated by the large number of names that it has in several languages, with Arabic, Indian, Sanskrit, Greek and Latin roots. It has many local names (see Chapter 2).

Uses

Almost all the species of the genus *Trigonella* are strongly scented (Anonymous, 1994) and most of them are used as insect repellent (Chopra *et al.*, 1949; Duke, 1986) for the protection of grains, cloths, etc.; while the essential oils of some of them are a very valuable raw material for the perfumery (Fazli and Hardman, 1968).

Most of these species (*T. foenum-graecum, T. caerulea, T. corniculata, T. hamosa, T. balansae, T. laciniata, T. marginata, T. occulta, T. anguina, T. arabica, T. glabra, T. stelata, T. coerulenses, T. spinosa, T. sibthorpii, T. spicata*, etc.) are rich in protein, vitamins and amino acids (Hidvegi *et al.*, 1984), while the seeds and the fresh material are used as forage, especially for cattle, mainly in the eastern Mediterranean area. In particulars *T. arabica* and *T. stelata* are foraged by animals in the desert areas of the Sahara, Palestine and the Dead Sea (Allen and Allen, 1981).

Several species of *Trigonella* (*T. foenum-graecum, T. balansae, T. corniculata, T. maritima, T. spicata, T. coerulea, T. occulta, T. polycerata, T. calliceras, T. cretica*, etc.) contain some interesting, from the pharmaceutical point of view, phytochemical compounds belonging to steroids, flavonoids and alkaloids (Anonymous, 1994) and efforts are being made to use some of them as a source of these constituents, especially of the steroidal diosgenin (Hardman, 1969). Seeds of these species also yield choline, a semicrystalline white saponin, a lactation-stimulating oil and various gums (Allen and Allen, 1981).

The alkaloid trigonelline has been isolated from plant parts, mainly seeds of *T. caerulea, T. cretica, T. foenum-graecum, T. lilacina, T. radiata, T. spinosa* (Allen and Allen, 1981) and *T. polycerata* (Mehra *et al.*, 1996). This pyridine alkaloid is known for its hypoglycemic and hypocholesterolaimic properties (Mehra *et al.*, 1996).

Some of these species are also used in traditional as well as veterinary medicine for different diseases, alone or in combination with other remedies: *T. occulta, T. polycerata* and *T. uncata* are included among the Indian herbals along with *T. foenum-graecum* (Hardman and Fazli, 1972).

The well developed endosperm of most of the species is rich in the polysaccharide mucilage (galactomannan) that has wide uses in industry including in pharmaceuticals and cosmetics.

In some parts of Pakistan and India *T. corniculata* is used for different purposes: its young tops are currently used as a green vegetable, the dried herb as a flavouring agent and its seeds for the treatment of swellings and bruises (Hardman and Fazli, 1972).

Chopped foliage of the species *T. caerulea* (sweet trefoil) is used in Switzerland for flavouring green cheeses: *Schabzieger, Chapsiger* and *Serred Vert*. In some parts of Tirol sweet trefoil is used for flavouring the bread called *Brotuⁿrze*. Sweet trefoil is also employed as a condiment in soups and potatoes, as a decoction for tea, and as flavouring in Chinese tea (Allen and Allen, 1981). Hardman and Fazli (1972) report that in Switzerland sweet trefoil has also been used in herbal medicine.

The varied and numerous special uses of the species *T. foenum-graecum* (fenugreek) are described in more detail in Chapter 4.

Need for research

This section reports on *T. foenum-graecum* (fenugreek), which is the only widely cultivated species of the genus *Trigonella*.

Fenugreek faces problems that keep it from reaching its full potential. Recently Edison (1995) reported that in India there are problems in improving the productivity of spices, one of which is fenugreek, due to:

- lack of advanced breeding methods for creation of high yielding varieties
- inherent inability expressed through poor and slow germination
- lack of adequate genetic variability
- lack of research based on crop rotation and cropping system
- inadequate techniques for diagnostic tests and screening for host resistance
- poor methods of nutrition and general management, particularly in light and sandy soils
- lack of incentives for seed production and poor storage facilities
- inadequate production and delivery systems of high quality planting material
- lack of facilitation of import genetic material for evolving new and improved varieties.

In order to overcome these problems, the following strategies have been under consideration (Edison, 1995):

- investigation of yield and loss forecasting models for both the producer and the trader
- import/exchange of valuable germplasm and promising varieties from the main regions of the Mediterranean in order to overcome the yield barrier
- production, distribution and delivery guaranteed planting material (certified seeds)
- identification on the basis of region suitable variety and choosing the best one together with the package of practices
- organization of demonstration fields
- motivating farmers to apply improved management techniques
- organizing cooperative markets and conducting producer–buyer–trader meetings in respective centres.

In India, in the last fifty years, eight research and development plans have been established for spices, including fenugreek, through a wide network of research institutions and state universities under the All India Coordinated Research Project (A.I.C.R.P, Edison, 1995). Similar problems are faced by fenugreek growers in other fenugreek producing countries throughout the world. Further, the necessary research information is not available to help them make correct decisions regarding existing problems.

It is a safe assumption, however, that all these problems can be solved by approaches through a well planned research programme taking into consideration the research priorities for fenugreek.

Projections must relate to comparative high yields, lower production costs, development of improved and better adapted varieties characterized by higher quantity and better quality, investigation of technological changes in production and utilization techniques and development of improved management practices; in general, fenugreek is grown under poor management conditions (Paroda and Karwasra, 1975).

A significant increase in yields through the suitable use of irrigation and adequate levels of soil fertility could make an immediate and important contribution to farm income. The very high yields recorded under experimental conditions (Petropoulos, 1973; Evans, 1989) and the reported big differences in seed yield among twenty-nine ecotypes (Banyai, 1973) suggest that it is not taking full advantage of the yield capacity of many fenugreek varieties.

Production cost is increasing and research must help the farmer, so that the money invested in increasing crop yields is reflected in the amount and the quality of collected seed or forage. Adaptable and improved genotypes and varieties are needed, suitable for: mechanization, no scattering of seed, high yielding and seed content characterized by high active constituents (diosgenin, protein etc., Cornish *et al.*, 1983), resistant to diseases, pests and drought. However, fenugreek is generally considered an unpretentious plant and rarely subject to diseases and pests (Sinskaya, 1961; Hardman, 1969; Duke, 1986).

The creation of a genotype without the peculiar smell that causes the tainting of animal products (milk, meat) and its derivatives (Molfino, 1947; Talelis, 1967) for an unlimited parallel use as forage for better valorization of the crop, should be another research objective. This objective should be based on the condition that progress in this goal is not offset by losses in some other valuable crop attributes.

There is also a need for research in the investigation and adaptation of new, more rapid and accurate analytical methods, for isolation and characterization of steroids, for analysis and utilization of the flavour extracts, the nutritive value of protein, the bread making ability of seed, and in general for the analysis and utilization of the other active constituents of fenugreek. The increase of the diosgenin content during the growing period through fertilization (Kozlowski *et al.*, 1982), use of herbicides (Mohamed, 1983) and other cultivation methods, as well as post harvest treatments through fermentation (Evans, 1989), incubation (Elujoba and Hardman, 1985), enzymes (Elujoba and Hardman, 1987), hormonal influence (Hardman and Stevens, 1978), tissue culture (Stevens and Hardman, 1974) and other biotechnological methods are some of the other critical areas.

The identification of the mechanisms of fenugreek galactomannan biosynthesis (during seed development) and hydrolysis (during germination) in order to produce transformed fenugreek plants, where the ratio Gal./Man. is appropriate for industrial use (Reid and Meier, 1970; Li *et al.*, 1980), needs further research efforts. The complete mechanization of sowing, harvesting, threshing and cleaning of fenugreek seed to increase yields and reduce the cost of production are also critical areas for research. This will help scientists to develop, through integrated research management programmes, means to establish optimum levels of fenugreek production and to optimize the yield of active constituents per unit area for a wide range of environmental and other conditions and for specific farming situations.

Future trends

The usefulness of fenugreek as a commercial and chemurgic crop is now being recognized, not only as a break-crop for cereal areas, where it is a very good soil renovator (Duke, 1986), but as forage, medicinal plant, source of diosgenin (the most important raw material for the steroid industry) and other constituents (protein, fixed oils, mucilage), as well as for culinary uses: as a traditional and modern flavouring.

Fenugreek is grown in about 57,000 ha with a production of 68,000 tons. Higher seed yield per hectare will be obtained through superior varieties and better management practices and may contribute to an increase in the crop worldwide; however, in India during the eighth plan of research and development, the overall growth rate of spices, including fenugreek, was 8 per cent. Fenugreek with the other spices, is a major source of foreign exchange for India (Edison, 1995).

From the world production of fenugreek it can be estimated that more than half is produced in India. India consumes domestically 90 per cent of its own production and claims 70–80 per cent of the world exports in fenugreek (Edison, 1995). Although the market for fenugreek is considerably small, there is a world incremental growth rate in demand of 4 per cent and a steady increase in exports as recently reported by Edison (1995), who is the key-man for spices of India and especially for fenugreek seeds, and later by an anonymous study (1996).

It is obvious that all this information on the characteristics and cultivation of fenugreek, like other specific crop plants, cannot serve as the sole basis for predicting immediate and long-term trends. But it is believed that the following facts, which have been noticed during the last years, open new prospects that could presage changes in farm practices that will affect positively the future of fenugreek production, especially in traditional fenugreek producing countries:

1   As recently reported by Edison (1995) there is an emerging change in food habits, preference for natural colouring and flavouring in fast food restaurants chains, microwave cooking, revolution in packaging and demand for quality assurance in relation to ISO 9000. It has been estimated that these changes will increase the world demand for spices, including fenugreek. Its exports from India increased in 1995–96, more than two-fold on an average over the last three years, reaching the amount of 15,135 tons (Anonymous, 1996). This increase in combination with the recently signed Uruguay Round Agreements for spice marketing (Nandakumar, 1997) will create new prospects for its cultivation.

2   One possible application, for which it is claimed fenugreek has good prospects, is its utilization as source of diosgenin, a steroidal precursor. Diosgenin is of importance to the pharmaceutical industry as a starting material in the partial synthesis of corticosteroids, sex hormones and oral contraceptives. At present, natural diosgenin is obtained mainly from the tubers of certain wild species of *Dioscorea* in Mexico, a process that is costly and difficult, requiring several years before the tubers grow to a size with significant content of diosgenin. On the other hand fenugreek is an unpretentious plant (Sinskaya, 1961; Hardman, 1969) and gives a consistent seed yield in a short growing period. The extraction of diosgenin from fenugreek may become attractive as today's widely used synthetic diosgenin will be implicated for some side-effects. But this extraction must be economically viable not through a fall in the price, but by the increase of its diosgenin content with genetic, agronomic and biotechnological methods and by reduction of the cost production, in such a manner that it will be attractive and be able to offer more prospects to growers.

3   The recent investigation of a technical development involving the spraying of liquid flavours of fenugreek on foodstuffs is claimed to give a better flavour dispersion than the usual method of simply sprinkling the dry flavour compound on the feed (Smith, 1982) and because of this the demand for fenugreek will increase rapidly. It is well known that the fenugreek flavour extract for animal feed, for both ruminants and pigs, is the main reason for fenugreek seed exports in the United Kingdom and other European countries (Smith, 1982).

4   The preparation of bread from fenugreek for those suffering from diabetes is ideal. It has less starch and polysaccharides are present in the form of silico-phosphoric ester of manogalactan, which is not hydrolyzed by ptyalin or pancreatic amylase (Kamel, 1932), and fenugreek seeds have an insulin stimulating substance (Hillaire-Buys *et al.*, 1993), plus a high protein content. This could be combined with the confirmed results during the last years of anti-diabetic (Sharma *et al.*, 1996) and anti-hypercholestrolaimic (Khosla *et al.*, 1995) effect of fenugreek seeds. This use is expected to seriously increase the consumption of fenugreek seed and to create better prospects for this crop in the future.

5   Due to the increasing protein deficiency all over the world, considerable efforts are being made to discover the nutritional potential of neglected sources. Thus, the aim now and even more in the future, is to utilize every protein source wherever and however it will have the highest nutritional value. Fenugreek protein is rich in lysine (345 mg g^{-1}) and in

comparison to the data for human requirements its quality, calculated from the amino-acid pattern, approaches that of the soybean (Hidvegi *et al.*, 1984). Therefore, it is an important crop for those countries in the Middle and Far East where meatless diets are customary for cultural and religious reasons.

The conclusion drawn is that the future of fenugreek is promising and its seed, as a source of diosgenin, which is the base for the production of oral contraceptives and rich in protein and fixed oils, could make a two-fold economic contribution to the world increase of population problems by assisting in birth control and at the same time providing additional food. The obvious growth in human population, due mainly to the increase in average life expectancy in the world because of the reduction in infant mortality, the progress in medicine and the improvement of food and residence conditions, results in pressures for human foods that will increase continuously. Fenugreek will have an important role to play, as many countries, especially in Asia and Africa, have fantastic opportunities to increase its production with no serious inroads on their supplies of cereal grains, for which fenugreek is a very good soil renovator (Duke, 1986). Therefore, population growth control can be achieved, further on a planet where the human population consumes the total production from every acre of tillable land additional food will be obtained.

References

Allen, O.N. and Allen, E.K. (1981) *The Leguminosae*, Macmillan Co., London.

Anonymous (1994) *Plants and Their Constituents, Phytochemical Dictionary of the Leguminosae*, Vol. 1, Cherman and Hall, London.

Anonymous (1996) *Spices Statistics*, Spices Board, Ministry of Commerce, Governement of India, P.B. No. 2277, Cochin.

Banyai, L. (1973) Botanical and qualitative studies on ecotypes of fenugreek (*Trigonella foenum-graecum* L.). *Agrobotanica*, **15**, 175–87.

Chopra, R.N., Badhwar, R.L. and Ghosh, S. (1965) *Poisonous Plants of India*, Vol. 1, Indian Council of Agricultural Research, New Delhi.

Cornish, M.A., Hardman, R. and Sadler, R.M. (1983) Hybridization for genetic improvement in the yield of diosgenin from fenugreek seed. *Planta Medica*, **48**, 149–52.

Duke, A.J. (1986) *Handbook of Legumes of World Economic Importance*, Plemus Press, New York and London.

Edison, S. (1995) Spices – research support to productivity. In N. Ravi (ed.), *The Hindu Survey of Indian Agriculture*, Kasturi & Sons Ltd., National Press, Madras, pp. 101–5.

Elujoba, A.A. and Hardman, R. (1985) Incubation conditions for fenugreek whole seed. *Planta Medica*, **51**(2), 113–15.

Elujoba, A.A. and Hardman, R. (1987) Saponin hydrolyzing enzymes from fenugreek seed. *Fitoterapia*, **58**(3), 197–9.

Evans, W.C. (1989) *Trease and Evans Pharmacognosy*, 13th edn, Balliere Tindall, London.

Fazli, F.R.Y. and Hardman, R. (1968) The spice fenugreek (*Trigonella foenum-graecum* L.). Its commercial varieties of seed as a source of diosgenin. *Trop. Sci.*, **10**, 66–78.

Furry, A. (1950) Les cahiers de la recherche agronomique. **3**, 25–317.

Hardman, R. (1969) Pharmaceutical products from plant steroids. *Trop. Sci.*, **11**, 196–222.

Hardman, R. and Fazli, F.R.Y. (1972) Methods of screening the genus *Trigonella* for steroidal sapogenin. *Planta Medica*, **21**, 131–8.

Hardman, R. and Stevens, R.G. (1978) The influence of N.A.A. and 2,4 D on the steroidal fractions of *Trigonella foenum-graecum* static cultures. *Planta Medica*, **34**, 414–19.

Hidvegi, M., El-Kady, A., Lásztity, R., Bekes, F. and Simon-Sarkadi, L. (1984) Contribution to the nutritional characterization of fenugreek (*Trigonella foenum-graecum* L.). *Acta Alimentaria*, **13**(4), 315–24.

Hillaire-Buys, D., Petit, P., Manteghetti, M., Baissac, Y., Sauvaire, Y. and Ribes, G. (1993) A recently identified substance extracted from fenugreek seeds, stimulates insulin secretion in rat. *Diabetologia*, 36, A 119.

Howard, M. (1987) *Traditional Folk Remedies. A Comprehensive Herbal*, Century Hutchinson Ltd., London.

Ivimey-Cook, R.B. (1968) *Trigonella* L. In T.G. Tutin, V.H. Heywood, N.A. Burges, D.M. Moore, D.H. Valentine, S.M. Walters, and D.A. Webb (eds), *Flora Europaea – Rosaceae to Umbelliferae*, Cambridge University Press, Cambridge 2, 150–2.

Jones, C.P. (1989) *Extracts from Nature*, Marks and Spencer P.L.C., Tigerprint, London.

Kamel, M.D. (1932) Reserve polysaccharide of the seeds of fenugreek. Its digestibility and its fat during germination. *Biochem. J.*, 26, 255–63.

Khosla, P., Gupta, D.D. and Nagpal, R.K. (1995) Effect of *Trigonella foenum-graecum* (fenugreek) on serum lipids in normal and diabetic rats. *Indian J. Pharmacol.*, 27, 89–93.

Kozlowski, J., Nowak, A. and Krajewska, A. (1982) Effects of fertilizer rates and ratios on the mucilage value and diosgenin yield of fenugreek. *Herba Polonica*, 28(3–4), 159–70.

Leyel, C.F. (1987) *Elixirs of Life*, Faber & Faber, London.

Li, X., Farn, M.-J., Feng, L.-B., Shan, X.-Q. and Feng, Y.-H. (1980) Analysis of the galactomannan gums in 24 seeds of *Leguminosae*. *Chin. Wu, Hsueh Pao*, 22(3), 302–4.

Lust, J.B. (1986) *The Herb Book*, Bantam Books Inc., New York.

Manniche, L. (1989) *An Ancient Egyptian Herbal*, British Museum Publ. Ltd., London.

Mehra, P., Yadar, R. and Kamal, R. (1996) Influence of nicotinic acid on production of trigonelline from *Trigonella polycerata* tissue culture. *Indian J. Experim. Biol.*, 34(11), 1147–9.

Melius, P. (1578) *Herbarium*, Heltai Gásparne Könyvnyomdája, Kolozsvár.

Miller, J.I. (1969) *The Spice Trade of the Roman Empire 29 B.C. to A.D. 641*, Clarendon Press, Oxford.

Mohamed, E.S.S. (1983) *Herbicides in Fenugreek (Trigonella foenum-graecum L.) with Particular Reference to Diosgenin and Protein Yields*, PhD Thesis, Bath University, England.

Molfino, R.H. (1947) Argentine plants producing changes in the characteristics of milk and its derivatives. *Rev. Farm.* (Buenos Aires), 89, 7–17.

Nandakumar, T. (1997) International spice marketing and the Uruguay Round Agreements. *International Trade Forum*, 1, 18–27.

Paroda, R.S. and Karwasra, R.R. (1975) Prediction through genotype environment interactions in fenugreek. *Forage Res.*, 1(1), 31–9.

Petropoulos, G.A. (1973) *Agronomic, genetic and chemical studies of Trigonella foenum-graecum L.*, PhD. Thesis, Bath University, England.

Reid, J.S.G. and Meier, H. (1970) Chemotaxonomic aspects of the reserve galactomannan in leguminous seeds. *Z. Pflanzenphysiol.*, 62, 89–92.

Rosengarten, F. (1969) *The Book of Spices*, Livingston, Wynnewood, Penns., USA.

Rouk, H.F. and Mangesha, H. (1963) *Fenugreek (Trigonella foenum-graecum L.). Its relationship, geography and economic importance*, Exper. Stat. Bull. No. 20, Imper. Ethiopian College of Agric. & Mech. Arts.

Schauenberg, P. and Paris, F. (1990) *Guide to Medicinal Plants*, Lutterworth Press, Cambridge, UK

Sharma, R.D., Sarkar, A., Hazra, D.K., Misra, I., Singh, J.B. and Maheshwari, B.B. (1996) Toxicological evaluation fenugreek seeds: a long term feeding experiment in diabetic patients. *Phytotherapy Research*, 10(6), 519–20.

Sinskaya, E. (1961) *Flora of cultivated plants of the U.S.S.R. XIII. Perennial leguminous plants*, Part I. *Medic, Sweet clover, Fenugreek*, Israel Programme for Scientific Translations, Jerusalem.

Smith, A. (1982) *Selected Markets for Turmeric, Coriander, Cumin and Fenugreek seed and Curry Powder*, Tropical Product Institute, Publication No. G 165, London.

Stevens, R.G. and Hardman, R. (1974) Steroid studies with tissue cultures of *Trigonella foenum-graecum* L. using G.L.C. *Proc. 3rd Intern. Congress of Plant Tissue and Cell Culture*, Leicester, 1974.

Stuart, M. (1986) *The Encyclopaedia of Herbs and Herbalism*, Orbis, London.

Talelis, D. (1967) *Cultivation of Legumes*, Agric. College of Athens, Athens (in greek).

Uphof, J.C.T. (1968) *Dictionary of Economic Plants*, Lehre Verlag von J. Cramer, New York.

2 Botany

Georgios A. Petropoulos and Panagiotis Kouloumbis

The genus *Trigonella*

Taxonomy

The genus *Trigonella* according to Hutchinson (1964) is one of the six genera (the other five are: *Parochetus*, *Melilotus*, *Factorovekya*, *Medicago* and *Trifolium*) of the Subfamily or Tribe *Trifoliae* of the Family *Fabaceae* (*Papilionaceae*) within the order *Leguminosae* (*Leguminales*).

Several investigators have attempted to employ the taxonomy of the genus *Trigonella*. Sirjaev (1933) has given in Latin an elaborate and systematic account of its taxonomy. Vasil'chenko (1953) has published a synopsis in Russian discussing the position of the genus within the Family *Leguminosae* and gave keys, synonyms and descriptions of the morphological characters of different series, their economic importance and geographical distribution. Hutchinson (1964), Heywood (1967) and Sinskaya (1961) have also given detailed descriptions of its taxonomic characters. According to these authors, the genus *Trigonella* contains mostly annual or perennial plants that are often strongly scented, and are described in the following terms.

Leaves pinnately 3-foliate; stipules adulate to the petiole; leaflets usually toothed and nerves often running out into teeth; flowers solitary or sessile or pedunculate in axillary heads or in short racemes; calyx teeth equal or unequal; corolla yellow, blue or purplish, free from the staminal tube or with wings united with prongs at the keel. Keel obtuse, shorter than the wings; stamens diadelphous or monadelphous with filaments not broadened; anthers uniform; stigma terminal; ovary sessile, ovules numerous. Pods varying greatly in size, cylindrical or compressed, linear or oblong, straight or curved, indehiscent or dehiscing with a pronounced short or long mucro (beak). Seeds, 1-many, finely or fairly markedly tuberculate, smooth; cotyledodns geniculate.

There is a big controversy about the number of species that comprise the genus *Trigonella*. Two hundred and sixty (260) species (182 from Linnaeus to 1885 and 78 from 1886 to 1965) are listed under this genus, but a close scrutiny reveals about ninety-seven distinct species (Fazli, 1967), while Vasil'chenko (1953) has described 128 species. Hector (1936), Kavadas (1956), Rouk and Mangesha (1963) and Hutchinson (1964), have reported about seventy.

The most interesting species of the genus *Trigonella* are presented in Table 2.1.

The reference to Index Kewensis (Hocker and Jackson, 1955) shows that much synonymity has occurred within the species of the genus *Trigonella*, that is, as has been reported in the section on Fenugreek, three different species have been described as *T. foenum-graecum*.

Table 2.1 A list of the well known species of the genus *Trigonella*[a]

T. anguina Del.	*T. marginata* Hochst. & Steud.
T. arabica Del.	*T. maritima* Poiret or Delile ex Poiret in Lam.
T. arcuata C.A. Mey	*T. melilotus caeruleus* (L.) Ascherson & Graebner[c]
T. aristata Vass.	*T. monantha* C.A. Mey
T. auradiaca Boiss. (= *T. aurantiaca* Boiss.)	*T. monspeliaca* L. (=*T. monspeliana* L.)[d]
T. balansae Boiss. and Reut. in Boiss.	*T. noδana* Boiss.
(=*T. corniculata* L.)	*T. occulta* Ser. Del.
T. berythaea Boiss. and Blanche	*T. ornithopoides* (L.) DC.[e]
T. brachycarpa (Fisch) Moris	*T. orthoceras* Kar. & Kir.
T. caelesyriaca Boiss.	*T. pamirica Gross.* in Kom.
T. caerulea (L.) Ser. (=*T. coerulea* L.)	*T. platycarpos* L.
T. calliceras Fisch ex Bieb.	*T. polycerata* L.
T. cancellata Dest.	*T. popovii* Kor.
T. cariensis Boiss.	*T. procumbens* (Besser) Reichenb.
T. coerulescens (Bieb.) Halacsy Hal.	*T. radiata* Boiss.
T. corniculata (L.) L. (=*T. balansae*	*T. rechingeri* Sirj.
Boiss. & Reut.)	*T. rigida* Boiss. & Bal.
T. cretica (L.) Boiss.[b]	*T. ruthenica* L.
T. cylindracea Desv. (=*T. culindracea* Desv.)	*T. schlumbergeri* Buser (Boiss.)
T. emodi Benth.	*T. sibthorpii* Boiss.
T. erata	*T. smyrnaea* Boiss.
T. fischeriana Ser.	*T. spicata* Sibth. an Sm. (=*T. homosa* Bess.)
T. foenum-graecum L.	*T. spinosa* L.
T. geminiflora Bunge	*T. sprunerana* Boiss. (=*T. spruneriana* Boiss.)
T. gladiata Stev. or Stev. ex Bieb.	(=*T. tortulosa* Gris.)
T. graeca (Boiss. and Spruner) Boiss.	*T. stellata* Forssk.
T. grandiflora Bunge	*T. striata* L.
T. hamosa L.	*T. suavissima* Lindl.
T. hybrida Pourr.	*T. tenuis* Fisch ex Bieb.
T. incisa Benth.	*T. tortulosa* Gris. (=*T. sprunerana* or *spruneriana* Boiss.)
T. kotschyi Fenzl. ex Boiss.	*T. uncata* Boiss. & Noe. (=*T. glabra* subs. *uncata*
T. laciniata (L.) Desf.	(Boiss. & Noe.) Lassen)
T. lilacina Boiss.	

Notes

a The botanical names have been completed according to the Index Kewensis (Hocker and Jackson, 1955).

b It has transformed to the genus *Melilotus* under the name *M. creticus*.

c It has fused with the species *T. caerulea* under the name *T. caerulea*.

d It has transformed to the genus *Medicago* under the name *M. mospeliaca* or *monspeliana*.

e It has transformed to the genus *Trifolium* under the name *T. ornithopoides*.

Further, in the Index Kewensis the following thirteen synonyms are given for the genus *Trigonella*:

1	*Aporathus*	Broamf.	(1856)
2	*Botryolotus*	Jaub	(1842)
3	*Buceras*[1]	Hall	(1785)
4	*Falcatula*	Brot	(1801)
5	*Foenum-graecum*	(Tourn) Rupp.	(1745)

1 Probably from the *Βουκέρας* (βούς = ox and *Κέρας* = horn) one ancient Greek name that Theophrastus had given for fenugreek.

6	*Follicullicera*	Pasq.	(1867)
7	*Grammocarpus*	Schur.	(1853)
8	*Kentia*	Adans	(1763)
9	*Melisitus*	Medic	(1787)
10	*Nephromedia*	Kostel	(1844)
11	*Pocockia*	Ser	(1825)
12	*Tellis*[1]	Linn.Syst.ed.I	(1735)
13	*Trifoliastrum*	Moench	(1794)

Some explanation for the assignment, reassignment and regroup of certain species between the genus *Trigonella, Medicago* and *Melilotus* is required. Brenac and Sauvaire (1996) proposed that pollinastanol and steroidal sapogenins should be used as chemotaxonomic markers to investigate the generic separation between the three genera. Their results support the unchanged assignment of *T. corniculata, T. caerulea* and *T. melilotus caeruleus*. They confirm the regroup of the last two species under the name *T. caerulea* and also the transform of *T. monspeliaca* to the genus *Medicago*. However, their results do not completely support the unchanged assignment of *T. calliceras* to the genus *Trigonella*, nor the reassignment of the *T. cretica* to *Melilotus cretica*, as the composition of this species is close to that of *T. foenum-graecum*, for the compounds investigated. The taxonomic transfer of *T. ornithopoides* (L.) DC. to the genus *Trifolium* appears justified in the light of rhizobial kinships (Allen and Allen, 1981). Also, the ratio Gal./Man. of the reserve galactomannan of the seed possesses a relative chemotaxonomical value as it varies among the different plant genus of *Leguminosae* (Reid and Meier, 1970).

According to Darlington and Wylie (1945) the chromosome contents for the genus indicate a basic haploid number of 8, 9, 11 and 14. Most of the species reported are diploid with 16 chromosomes. However *T. homosa* from Egypt is reported to have 16 and 44 chromosomes, *T. ornithoides* from Europe 18, and *T. polycerata* from the Mediterranean and South West Asia 28, 30 and 32.

Tutin and Heywood (1964) divide the genus *Trigonella* into three subgenera, according to the form and shape of the calyx and pod, as follows:

a   Subgenus *Trigonella*: Calyx usually campanulate. Pod not inflated, with representatives of the species *T. graeca, T. cretica, T. maritima, T. corniculata*.

b   Subgenus *Trifoliastrum*: Calyx campanulate. Pod inflated with representatives of the species *T. caerulea* and *T. procumbes*.

c   Subgenus *Foenum-graecum*: Calyx tubular. Pod not inflated with representatives of the species *T. foenum-graecum* and *T. coerulescens*.

Ingham (1981) found that three groups of species occur in *Trigonella*, based on results of their ability to release coumarin on tissue maceration. Two of these groups linking the genus *Medicago, Factorovekya* and *Melilotus* and the third group with the genus *Trifolium*.

Furry (1950) also divided the cultivated species of the genus *Trigonella*, according to the colour of the corolla and other characters, as follows:

a   Corolla blue: *T. caerulea*

b   Corolla whitish: *T. foenum-graecum*

c   Corolla yellow:

   i   Plant annual, calyx with teeth equal to the tube: *T. polycerata*

   ii   Plant annual, calyx with teeth longer than the tube: *T. monspeliaca*

   iii   Plant perennial: *T. suavissima*

We do not agree completely with the corolla colour of the species *T. foenum-graecum* reported above, as in our experiments this colour was yellow from the beginning and for most of the flowering period and only at the end, if at all, did the colour turn whitish.

Distribution

The Mediterranean region is known to be the natural habitat of the genus *Trigonella*. Species of the genus exist wild in the countries of Europe, Macaronesia (Canarian Islands) North and South Africa, Central Asia and Australia (Anonymous, 1994).

Indigenous species of this genus have been reported (Anonymous, 1994): six for Asia (*T. caelesyriaca, T. calliceras, T. emodi, T. geminiflora, T. glabra, T. kotschyi*), five for Europe (*T. graeca, T. striata, T. polycerata, T. monspeliaca, T. procumbens*), one for Africa (*T. laciniata*) and one for Australia (*T. suavissima*), where it has adapted well to the wet swampy habitat (Allen and Allen, 1981). The rest of the species exist in more than one continent, that is, twenty-three species of this genus have been reported for Europe (Ivimey-Cook, 1968), of which fifteen occur in the Balkan area (Polunin, 1988) including the fourteen for Greece (Kavadas, 1956), of which four occur in the famous Island Kefallinia (Phitos and Damboldt, 1985).

However, the most interesting species of the genus is the widely cultivated *T. foenum-graecum* (fenugreek), which is described in detail.

Fenugreek (*T. foenum-graecum* L.)

Taxonomy

According to Sinskaya (1961), Hutchinson (1964), Tutin and Heywood (1964) and our observations the chief taxonomic characters of the species *T. foenum-graecum* are the following.

Stems 20–130 cm long, straight, rarely ascending, branching, rarely simple, sparsely pubescent, usually hollow, anthocyanin tinged at base or all the way up, rarely completely green. First leaf simple, some times weak trifoliate, oval or orbicular with entire margin and a long petiole. Stipules fairly large, covered with soft hair. Leaf petiole thickened at the top, attenuate beyond point of attachment of lateral leaflets. Petiolules very small cartilaginous. Petioles and petiolules vested on the underside with simple, soft sparse hairs. Leaflets from ovate-orbicular to oblong-lanceolate, 1–4 cm long, almost equal, finely haired, dentate, near the apex, dentation more strongly developed in upper than in lower leaves. The petioles and the blades of the leaflets are anthocyanin-tinged to a varying degree of green. Flowers in leaf axils, mostly twin, more rarely solitary (we distinguished the cleistogamy and aneictogamy type of flowers). Calyx 6–8 mm, soft hairy with teeth as long as the tube, half as long as the corolla. Corolla 13–19 mm long pale yellow (white at the end of flowering period), some times lilac coloured at the base. Standard tend backwards oblong emarginate at apex with bluish spots (these spots are absent from some genotypes), wings half as long as the standard; keel obtuse, split at base. Pods with the mucro (beak), 10–18 cm long and 3.5×5 cm broad, curved, rarely straight, with transient hairs. Before ripening the pod is green or reddish coloured; when ripe light straw or brown containing 10–20 seeds.

Seeds vary from rectangular to rounded in outline with a deep groove between the radicle and cotyledons, the length is 3.5–6 mm and the width 2.5–4 mm, light greyish, brown, olive green or cinnamon coloured, with a pronounced radicle that is half the length of the cotyledons.

The minute hilum lies partly obscured with a deep notch. Odour characteristic. Chromosome number, $2n = 16$.

Linnaeus (1737, 1753) have described the species *T. foenum-graecum* first. The botanical names and synonyms assigned to fenugreek according to the Index Kewensis (Hocker and Jackson, 1955) are as follows:

1	*Foenum-graecum*	Linn. sp. pl. 777 Eur. oriens
2	*Foenum-graecum*	(Tourn) Rupp. FL, Jen. Ed. Hall 263 (1745)
3	*Graeca*	St. Lag. in Ann. Soc. Bot. Lyon VII (1880)
4	*Hausknechtii*	(Siry) in obs. *T. foenum-graecum* var. *Hausknechtii* (1933)
5	*Tibetana*	(Alef) in obs. *T. foenum-graecum officinale* var. *tibetanum*
6	*Rhodantha*	(Alef) in obs. *T. foenum-graecum officinale* var. *rhodanthus*

Mathé (1975) gives the following synonyms for the species *T. foenum-graecum* (L.):

1. *Buceras foenum-graecum* (L.) All.
2. *Foenum-graecum sativum* Medik.
3. *Foenum-graecum officinale* Moench.
4. *Foenum-graecum officinale* ssp. *cultum* Alef.
5. *Folliculigera graveolens* Pasq.
6. *Medicago foenu-graeca* Ehz Krause.
7. *Telis foenum-graecum* (L.) O.ktze.
8. *Trigonella graeca* St.Lag. non Boiss.
9. *Trigonella ensifera* Trautv.

Hocker and Jackson (1955) also report three different species of *Trigonella* as having been described as *T. foenum-graecum*:

1. The species *T. gladiata* (Hall) Desc. 138
2. The species *T. cariensis* Sibth and Sm. Fl. Graec.VIII 48+ 766
3. The species *T. monspeliaca* Suter, Fl. Helv. ed. Hegetachw. II 149

Serpukhova (1934) on the basis of N.I. Vavilev's collection of fenugreek in Yemen and Abyssinia, divided the cultivated fenugreek by its whole plant characters into two subspecies:

a   *T. foenum-graecum* L. ssp. *iemensis* (referring to the Yemen), which she established, with short stems and flowers, entire marginate leaflets, lanceolate and short calyx teeth, erect standard with dots, dried corolla at base of pod, short and lanceolate pod, small number of leaves and short vegetation period.

b   *T. foenum-graecum* L. ssp. *culta* (Alefeld) Gams, which had been first noted by Fluckiger and Hanbury (1879), characterized by taller plants, with dentate leaflets, long flowers, subulate and long calyx teeth, reflexed and without dots standard, at end of break dried corolla, long and linear pod, many leaves and long vegetation period.

Serpukhova (1934) also showed the polymorphic character of fenugreek and studied its variability in detail.

Sinskaya (1961) divided *T. foenum-graecum* into series, subseries and ecotypes based upon the taxonomical characters of the plant and gave an account of the morphological characters and habits of each subspecies and ecotypes.

Also, fenugreek plants have been distinguished in *pallida* and *colorata* type and described in detail (Petropoulos, 1973).

Moschini (1958) divided the cultivated fenugreek in Italy into three ecotypes:

   i   *Sicilian*, characterized by high precocity and high yield
  ii   *Toscanian*, late in maturity, resistant to cold and high yielding
 iii   *Moroccan*, with high precocity, resistant to cold and low yielding

Serpukhova (1934) classified the seeds of *T. foenum-graecum* according to their shape, size and colour and distinguished three groups (*Indicae, Anatolicae* and *Aethiopicae*), with one variety for the groups *Indicae* (nano-fulva) and Anatolicae (magno-fulva) and six varieties for the group Aethiopicae (fulva, punctato-fulva, olivacea, punctato-olivacea, leucosperma and griseo-coerulescens), while Fazli and Hardman (1968) give one version of her classification. Sinskaya (1961) later confirmed Serpukhova's classification, although he preferred to use the term 'forms' rather than 'varieties'.

Furry (1950) also divides fenugreek seeds into six types (*Yemenese, Transcaucasian, African, Afghan, Chinese-Persian* and *Indian*) and gives details only for the African type, in which he distinguishes two varieties (*North African* and *Sudanese-Egyptian of Kharthoum*).

The seeds of a rich collection of fenugreek samples (more than 300) of Bath University, originated from the countries of its cultivation, by a careful examination of their general appearance and other characteristics and in association with the country of origin, can be classified into the following four types (Petropoulos, 1973):

1   *Fluorescent type*: Seeds fluorescent under UV light, absence of any pigment in its seed coat, large (5–6 × 3–4 mm) rounded in outline, with high, one thousand seed weight (27–32 g) and Germ./Husk. index, probably induced by spontaneous mutation from Ethiopian populations, as most of its characters are controlled by recessive genes, not described previously. It is easily identified. Representatives of this type are the breeding cultivar Fluorescent and the variety 'Barbara'.

2   *Ethiopian type*: Non fluorescent under UV light, moderate in size (4.0–4.5 × 3.0–3.5 mm) with at least four different pigments in its seed coat and a thousand seed weight 22–25 g. It is a natural mixture of Serpukhova's olivacea and punctato-olivacea. In this type belong most of the samples from Ethiopia and its neighbouring fenugreek producing countries. It is a uniform type and very easily distinguished. Representatives of this type are the seeds of the Ethiopian breeding cultivar.

3   *Indian type*: Non fluorescent under UV light, with at least four pigments in its seed coat, very small seeds (2.5–3.5 × 2.0–2.5 mm), rectangular in outline, nano-fulva according to Serpukhova's classification, a thousand seed weight 15–20 g. In this type belong most of the samples from India, Pakistan, China and Kenya, the latter being bigger than the rest. This is also a uniform type and very easily distinguished. Representatives of this type are the seeds of the Kenyan breeding cultivar.

4   *Mediterranean type*: Non fluorescent under UV light. Large seeds (4.5–6.0 × 3.5–5.0 mm), rectangular in outline, a thousand seed weight 25–31 g, a natural mixture of magno-fulva, fulva and punctato-fulva according to Serpukhova's classification. In this type belong samples from Israel where magno-fulva was dominant, from Morocco, Portugal, Spain and

France where the punctato-fulva was dominant and from Greece and Turkey where the fulva was dominant. It is the least uniform and is not easily identified. Representative of this type are seeds of the Moroccan breeding cultivar.

Distribution and vernacular names

The species *T. foenum-graecum*, wild or cultivated, is widely distributed throughout the world, as is indicated by the great number of names it possesses with Arabic, Indian (Sanskrit) and European (Greek and Latin) roots. Fenugreek has been reported as a cultivated crop in Portugal, Spain, United Kingdom, Germany, Austria, Switzerland, Greece, Turkey, Egypt, Sudan, Ethiopia, Kenya, Tanzania, Israel, Lebanon, Morocco, Tunisia, India, Pakistan, China, Japan, Russia, Argentine and USA (Rouk and Mangesha, 1963; Fazli and Hardman, 1968; Rosengarten, 1969; Smith, 1982; Edison, 1995).

The genetic name, *Trigonella*, comes from Latin meaning 'little triangle', in reference to the triangular shape of the small yellowish-white flowers. The species epithet *foenum-graecum* means 'Greek hay' and according to Rosengarten (1969) the Romans, who got the plant from Greece where it was a very common crop in ancient times, gave it this name. It is also called 'ox horn' or 'goat horn' because of the two seed pods projecting in opposite directions usually from the nodes of the stem base that resemble ox or goat horns.

The main national names for this species are listed in Table 2.2.

Table 2.2 Natural or local names of fenugreek, in different countries

Speaking language of country	National or local names of fenugreek
Arabic	Hhelbah, Hhelbeh, Hulba, Hulabah
Armenian	Shambala
Azerbaijani	Khil'be, Boil
Chinese	K'u-Tou
Croatic	Piskayika, ditelina rogata
Czech	Piskayika, recke seno
Dutch	Fenegriek
English	Fenugreek, fenigrec
Ethiopian	Abish
French	Fenugrec, Senegre
German	Griechisch Heu, Griechisches Heu, Bockshornklee, Kuhhornklee, Bisamklee
Greek (modern)	Trigoniskos (Τριγωνίσκος), Tsimeni (Τσιμένι), Tintelis (Τιντελίς), Moschositaro (Μοσχοσίταρο), tili (τήλι), tipilina (τηπιλίνα)
Greek (ancient)	βούκερας ο φαρμακευτικός, τῆλις
Hungarian	Görögszéna
Indian	Methi
Italian	Fieno greco
Japanese	Koroba
Pakistani	Methi
Persian (Irani)	Schemlit
Polish	Fengrek, Kozieradka
Portuguese	Alforva
Russian	Pazhitnik, Pazsitnyik, Grezsezki szeno (гре́ческий се́но)
Slovak	Seneyka grecka, seno grecka
Swedish	Bockhornsklover
Uzbekistani	Khul'ba, Ul'ba, Boidana

References

Allen, O.N. and Allen, E.K. (1981) *The Leguminosae*, Macmillan Co., London.

Anonymous (1994) *Plants and Their Constituents. Phytochemical Dictionary of the Leguminosae*, Vol. 1, Cherman & Hall, London.

Brenac, P. and Sauvaire, Y. (1996) Chemotaxonomic value of sterols and steroidal sapogenins in the genus *Trigonella*. *Biochem. Systemat. Ecol.*, **24**(2), 157–64.

Darlington, C.D. and Wylie, A.P. (1945) *Chromosome Atlas of Flowering Plants*, George Allen & Unwin Ltd., London.

Edison, S. (1995) Spices – research support to productivity. N. Ravi (ed.), *The Hindu Survey of Indian Agriculture*, Kasturi & Sons Ltd., National Press, Madras, pp. 101–5.

Fazli, F.R.Y. (1967) *Studies in steroid-yielding plants of the genus Trigonella*, PhD Thesis, University of Nottingham, England.

Fazli, F.R.Y. and Hardman, R. (1968) The spice fenugreek (*Trigonella foenum-graecum* L). Its commercial varieties of seed as a source of diosgenin. *Trop.Sci.*, **10**, 66–78.

Fluckiger, F.A. and Hanbury, D. (1879) *Pharmacographia*, Macmillan & Co., London.

Furry, A. (1950) Les cahiers de la recherche agronomique, **3**, 25–317.

Hector, J.N. (1936) *Introduction to the Botany of Field Crops (Non cereals)*, Central News Agency Ltd., Johannesburg.

Heywood, V.H. (1967) *Plant Taxonomy – Studies in Biology No. 5*, Edward Arnold Ltd.

Hocker, J.B. and Jackson, D. (1955) *Index Kewensis*, Tomus II, 1116–1117 (1895) Suppl. XII, 146 (1951–1955), Clarendon Press, Oxford.

Hutchinson, J. (1964) *The Genera of Flowering Plants*, Vol. 1, Clarendon Press, Oxford.

Ingham, J.L. (1981) Phytoalexin induction and its chemosystematic significance in the genus *Trigonella*. *Biochem. Systemat. Ecol.*, **9**(4), 275–81.

Ivimey-Cook, R.B. (1968) *Trigonella* L. In T.G. Tutin, V.H. Heywood, N.A. Burges, D.M. Moore, D.H. Valentine, S.M. Walters and D.A. Webb (eds), *Flora Europaea – Rosaceae to Umbelliferae*, Vol. 2, Cambridge University Press, Cambridge, pp. 150–2.

Kavadas, D.S. (1956) *Illustrated Botanical – Phytological Dictionary*, Vol. XIII, pp. 3929–33 (in greek).

Linnaeus, C. (1737) *General Edition*, I, 351, Stockholm.

Linnaeus, C. (1753) *Species Plantarum*, Silvius, Stockholm, p. 1200.

Máthé, I. (1975) *A görögszéna (Trigonella foenum-graecum* L.), *Magyarország* III/2, Kultúrflóra 39, Akadémiai Kiadó, Budapest.

Moschini, E. (1958) Charatteristiche biologiche e colturali di *Trigonella foenum-graecum* L. e di *Vicia sativa* L. di diversa provenienza. *Esperienze e Ricerche*, pp. 10–11, Pisa.

Petropoulos, G.A. (1973) *Agronomic, genetic and chemical studies of Trigonella foenum-graecum* L., PhD. Thesis, Bath University, England.

Phitos, D. and Damboldt, J. (1985) Die Flora der Insel Kefallinia (Griechenland). *Botanika Chronika*, **5**(1–2), 1–204.

Polunin, O. (1988) *Flowers of Greece and the Balkans. A Field Guide*, 1.Repr., Oxford University Press, Oxford, New York.

Reid, J.S.G. and Meier, H. (1970) Chemotaxonomic aspects of the reserve galactomannan in leguminous seeds. *Z. Pflanzenphysiol.*, **62**, 89–92.

Rosengarten, F. (1969) *The Book of Spices*, Livingston, Wynnewood, Penns., USA.

Rouk, H.F. and Mangesha, H. (1963) *Fenugreek (Trigonella foenum-graecum* L.). *Its relationship, geography and economic importance*, Exper. Stat. Bull. No. 20, Imper. Ethiopian College of Agric. & Mech. Arts.

Serpukhova, V.I. (1934) Trudy, *Prikl. Bot. Genet. i selekcii Sen.*, **7**(1), 69–106 (Russian).

Sinskaya, E. (1961) *Flora of Cultivated Plants of the U.S.S.R. XIII, Perennial Leguminous plants, Part I, Medic, Sweet Clover, Fenugreek*, Israel Programme for Scientific Translations, Jerusalem.

Sirjaev, G. (1933) *Generis Trigonella L. rivisio critica*, Publ. Fac. Sci. Univ. Masaryk Brno, pp. 124–269.

Smith, A. (1982) *Selected markets for turmeric, coriander, cumin and fenugreek seed and curry powder*, Tropical Product Institute, Publication No. G165, London.

Tutin, T.G. and Heywood, V.H. (1964) *Flora Europaea*, Vol. I and II, Cambridge University Press, Cambridge.

Vasil'chenko, I.T. (1953) Bericht uber die Arten der Gattung. *Trigonella Trudy Bot. Inst. Akad. Nauk. S.S.S.R.* 1, 10.

3 Physiology

Caroline G. Spyropoulos

Seed physiology

Seed structure and composition

Although there are as many as seventy-two *Trigonella* species, most studies on seed structure and physiology have been performed on the *Trigonella foenum-graecum* L. (fenugreek).

Fenugreek seeds are surrounded by the seed coat. The seed coat is separated from the embryo by a well developed endosperm, which is the principal storage organ. In mature seeds the majority of the endosperm cells are nonliving, the cytoplasmic contents of which are occluded by the store reserves: galactomannan. This tissue is surrounded by a one cell layer of living tissue: the aleurone layer. The aleurone layer cells are small and thick walled and contain aleurone grains, which disappear during the course of seed germination (Reid and Meier, 1972; Bewley and Black, 1994).

The role of endosperm galactomannan is dual: it serves as a reserve material that will support the seedling growth during the early post-germination phase, but also, due to its high water retention capacity regulates the water balance of the embryo during germination (Reid and Bewley, 1979).

The embryo, as in all dicotyledons, is composed of a cotyledon pair and the embryo axis. Apart from the endosperm reserves, there are also reserves in the embryo (proteins, lipids, sugars) that will be metabolised upon seed germination and will be used for the growth needs of the young seedling (Bewley *et al.*, 1993).

The fenugreek seed coat apart from its protective character seems also to play a regulatory role in the mobilisation of the endospermic food reserves (Spyropoulos and Reid, 1985; 1988; Zambou *et al.*, 1993; Kontos *et al.*, 1996).

Seed development

Seed development starts upon fertilisation of the egg cell in the embryo sac, by one of the male pollen tube nuclei, and the fusion of the two polar nuclei in the embryo sac with the other pollen tube nucleus. The result is the formation of the embryo and the endosperm, respectively. The fenugreek seed development lasts approximately 120 days after anthesis (DAA) (Campbell and Reid, 1982). Galactomannan accumulation in the endosperm starts approximately 30 DAA and ends at approximately 55 DAA, just before the seed's fresh weight starts decreasing (Campbell and Reid, 1982).

Galactomannan synthesis during seed development: morphology

Galactomannan is deposited as cell wall thickenings of the endosperm cells and its deposition continues until nearly all the cytoplasm disappears. The only endosperm cells that are not filled with galactomannan are the cells of the aleurone layer. In these cells some galactomannan is deposited only at the outer walls next to the seed coat, at the cell corners, and occasionally at the side walls (Meier and Reid, 1977).

Galactomannan is deposited first in those cells that are neighbouring the embryo, while in those next to the aleurone layer, is deposited at the end (Meier and Reid, 1977).

An electron microscopy examination of fenugreek endosperms during the course of galactomannan deposition suggests that galactomannan synthesis takes place in the rough endoplasmic reticulum, it is accumulated in the netlike enchylema space and released outside the plasmalemma without the participation of the Golgi apparatus (Meier and Reid, 1977).

Galactomannan synthesis during seed development: biosynthesis

The biochemistry of galactomannan synthesis and mobilisation has attracted much interest, not only due to its biological importance, but also due to galactomannan extensive application in industry, notably, food, pharmaceuticals, cosmetics, paper products, paints, plasters, etc. (Dea and Morrison, 1975; Reid, 1985; Scherbukhin and Anulov, 1999). The ratio of mannose to galactose varies in the different plant genus but the most appropriate for industry applications is 4:1. Among the eight *Trigonella* species studied, all have mannose: galactose ratio approximately 1:1; only *T. erata* has a ratio of 1.6:1 (Reid and Meier, 1970). Galactomannan biosynthesis has been studied using cell free extracts and whole endosperm tissue (Edwards *et al.*, 1989; 1992).

The synthesis of galactomannan *in vivo* started about thirty DAA and its deposition increased until fifty-five DAA. There was a parallel increase in the activities of the mannosyl- and galactosyltransferases. The galactomannan present at any time of seed development had a mannose to galactose ratio of 1:1, the same with that of mature seeds (Edwards *et al.*, 1992).

The enzymes responsible for fenugreek galactomannan biosynthesis were two membrane bound glycosyltransferases, a GDP-mannose-dependent mannosyltransferase and a UDP-galactose-dependent galactosyltransferase. The mannosyltransferase catalyses the addition of mannose residues onto an unknown endogenous primer, which could be galactomannan. The addition of galactose residues by the action of the galactosyltransferase takes place only on newly transferred mannose residues on the mannan backbone. The regulation of the mannose to galactose ratio of the galactomannan by fenugreek is regulated by the enzyme galactosyltranferase (Reid *et al.*, 1992; 1995). Recently, Edwards *et al.* (1999) isolated a 51 kDa protein, with galactosyltransferase activity and isolated and cloned the corresponding cDNA. This cDNA encodes a protein, with a single transmembrane α-helix near the N terminus, which proved to be galactosyltransferase.

The mechanisms that underlie fenugreek galactomannan biosynthesis could lead to the production of transformed fenugreek plants with the required ratio of mannose to galactose (i.e. 4:1), which is suitable for industrial applications.

Seed germination and endosperm reserve mobilisation

Fenugreek seeds germinated approximately 10 h after the start of seed imbibition at 25°C in the dark (Reid and Bewley, 1979; Spyropoulos and Reid, 1985). Endosperm galactomannan mobilisation started after about 15 h of imbibition (Reid, 1971; Spyropoulos and Reid, 1985) through the action of α-galactosidase (EC 3.2.1.22), endo-β-mannanase (EC 3.2.1.78), and exo-β-mannanase (EC 3.2.1.25) (Reid *et al.*, 1977; Meier and Reid, 1982; Reid, 1985). The first two enzymes seem to be synthesised *de novo* while the third one is present in an active state in the endosperm of the dry seed. A very low α-galactosidase activity was detected in the dry seed, which was suggested to be involved in the hydrolysis of the raffinose series oligosaccharides (Reid and Meier, 1972), while endo-β-mannanase activity was absent (Reid *et al.*, 1977; Spyropoulos and Reid, 1988). During the course of seed imbibition the activity of α-galactosidase increased. Endo-β-mannanase activity appeared after 20 h of imbibition and increased thereafter. The increase of the activities of both hydrolases coincides with the decrease in galactomannan content in the endosperm (Figure 3.1).

The ultimate products of galactomannan hydrolysis, D-galactose and D-mannose, do not accumulate in the endosperm. Both monosaccharides are transported immediately to the embryo by carriers that have high specificity for the corresponding sugars. These carriers seem to play an important role in the switching on and off the uptake capacity of these sugars by fenugreek embryo (Zambou and Spyropoulos, 1989; 1990). The inhibition of galactose uptake by cycloheximide may suggest that the galactose carrier is synthesised *de novo* during imbibition. Although galactose and mannose uptake by the embryo is under metabolic control, their uptake does not take place via a H^+ co-transport system. It has been speculated that the metabolic energy needed for their uptake is used for the phosphorylation of these sugars, thus ensuring their transformation in cotyledons and consequently the generation of a concentration gradient between the endosperm and cotyledons.

The disappearance of galactomannan from the embryo is concomitant with the appearance of transitory starch and high levels of sucrose in the embryo (Reid, 1971; Bewley *et al.*, 1993), which are formed by the galactomannan hydrolysis products taken up by the embryo. Although

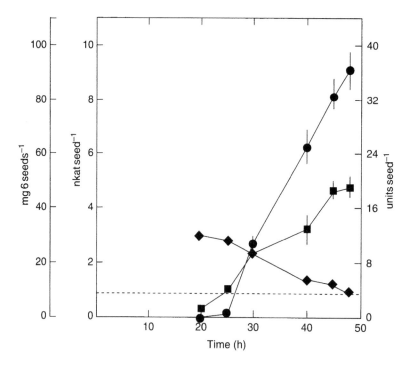

Figure 3.1 α-Galactosidase (■) and endo-β-mannanase (●) activities in the endosperms of fenugreek seeds and the dry weight of six extracted endosperms + testa (♦) at several imbibition times. The dashed line indicates the dry weight of testa, which does not change (Reid and Bewley, 1979). Decreases in dry weight are due to galactomannan mobilisation (Spyropoulos and Reid, 1988).

the initiation of starch formation in fenugreek cotyledons is independent of a supply of the galactomannan hydrolysates in the embryo, their presence is necessary for its accumulation (Bewley *et al.*, 1993).

There are several factors that regulate galactomannan mobilisation. The prerequisites for its mobilisation are:

1 The production of the enzymes that hydrolyse galactomannan, that is, α-galactosidase and endo-β-mannanase.
2 The secretion of these enzymes through the plasmalemma of the aleurone cells and their diffusion though the aleurone cell wall to reach their site of action.
3 The appropriate conditions for the action of these enzymes *in situ*.

Galactomannan mobilisation and the production of α-galactosidase and endo-β-mannanase may take place in isolated endosperms if they are incubated in a large volume under germination conditions (Reid and Meier, 1972; Spyropoulos and Reid, 1985; Malek and Bewley, 1991). In contrast, incubation of endosperms in a small volume resulted in the inhibition of α-galactosidase (Table 3.1) (Spyropoulos and Reid, 1985) and endo-β-mannanase production (Malek and Bewley, 1991; Kontos *et al.*, 1996). The effect of the small volume incubation medium was relieved if incubation was preceded by a 2-h-endosperm leaching, suggesting that in the endosperm and/or seed coat there are leachable inhibitory substances the diffusion of which is prevented when the volume of the incubation medium is small. Zambou *et al.* (1993) have isolated three substances from the leachate of fenugreek endosperm and seed coat, which inhibited the production of α-galactosidase by fenugreek endosperm and, chromatographically, behaved like saponins. These substances, however, did not have any effect on the production of these hydrolases if endosperms were treated after the start of the galactomannan mobilisation.

Removal of the embryo axis inhibited galactomannan mobilisation and the activity of α-galactosidase, suggesting that the embryo axis controlled galactomannan mobilisation. The effect of embryo axis excision on galactomannan hydrolysis and the activity of α-galactosidase was relieved upon addition of the excised axes into the 'seed' incubation medium or incubation of these 'seeds' with benzyladenine (BA) or BA plus GA_3 (Table 3.1). Initially, the axis appeared to have a regulatory function in determining the onset of α-galactosidase production in the endosperm. However, its continuous presence was necessary for the uptake of the galactomannan hydrolysis products, the accumulation of which inhibited galactomannan breakdown (Spyropoulos and Reid, 1985; 1988).

Table 3.1 α-Galactosidase activity and galactomannan levels in endosperms of fenugreek seeds after 48 h of imbibition, following the excision of the axis after 5 h

Incubated seed part	Incubation medium (volume ml)	α-Galactosidase nkat seed^{-1}	Galactomannan[a] mg 6 seeds^{-1}
Seed – axis	Water (0.5 ml)	0.35 ± 0.10	10.6 ± 1.8
Seed – axis	Water + excised axes (0.5 ml)	2.30 ± 0.13	7.7 ± 0.7
Seed – axis	10^{-5} M BA (0.5 ml)	1.86 ± 0.37	3.8 ± 0.5
Seed – axis	10^{-4} M GA_3 (0.5 ml)	2.02 ± 0.15	6.3 ± 0.4

Note
a Dry weight of six extracted endosperms + testae, less 9.0 mg, the average weight of the testae (Reid and Bewley, 1979).

Table 3.2 α-Galactosidase and endo-β-mannanase activity of leached or non-leached endosperms that were isolated from 5-h-imbibed seeds and incubated in water, in the endosperm + testa leachate or in PEG, for 20 h (Spyropoulos and Reid, 1988; Kontos *et al.*, 1996)

Treatment	α-Galactosidase nkat seed^{-1}	Endo-β-mannanase units endosperm^{-1}
Non-leached → water		8
Leached 2 h → water	0.9	14
Leached 2 h → leachate	0.15	0.3
Non-leached → PEG[a]	0.2	2.9
Leached 2 h → PEG[a]	0.7	7.6

Note
a PEG 3350, −1.5 MPa.

Water stress inhibited galactomannan mobilisation. When water stress was imposed on isolated endosperms before the onset of galactomannan breakdown there was a total inhibition of the production of α-galactosidase and endo-β-mannanase and consequently galactomannan hydrolysis did not take place (Table 3.2) (Spyropoulos and Reid, 1988; Zambou *et al.*, 1993; Kontos *et al.*, 1996). However, if water stress treatment on fenugreek endosperms was preceded by a 2-h-leaching, the effect of water stress on the production of the two hydrolases was repaired (Table 3.2) (Spyropoulos and Reid, 1988). These results suggested that under water stress conditions the removal of the endosperm and seed coat inhibitory substances was prevented.

When water stress was imposed after the start of galactomannan breakdown (on 25-h-imbibed seeds), although the production of both hydrolases was not affected, galactomannan breakdown was still inhibited (Spyropoulos and Reid, 1988). The inhibition of galactomannan breakdown could be attributed to either the inhibition of the galactomannan hydrolases secretion and/or their diffusion through the aleurone cell wall or to the inhibition of α-galactosidase action *in situ*.

Carob (*Ceratonia siliqua*) endosperm is a galactomannan reserving tissue (Seiler, 1979; Spyropoulos and Lambiris, 1980). Water stress imposed on carob endosperm protoplasts did not affect the production of α-galactosidase or endo-β-mannanase nor their secretion. However, experiments performed with whole carob endosperms have shown that under water stress conditions, the diffusion of these hydrolases into the endosperm incubation medium was inhibited. These results suggest that the carob endosperm cell wall controls galactomannan hydrolysis by the regulation of the diffusion of galactomannan hydrolases to reach the site of their action (Kontos and Spyropoulos, 1995). Likewise, it could be postulated that water stress affects the cell wall porosity of the fenugreek aleurone layer resulting in the decreased diffusion of the galactomannan hydrolysing enzymes.

Under water stress conditions, the amount of galactose taken up by the embryo was reduced because under these conditions the galactose carrier did not function (Zambou and Spyropoulos, 1990). Therefore, most galactose produced through the action of α-galactosidase would remain in the endosperm. Galactose is a potent inhibitor of α-galactosidase and its presence inhibits its action *in situ* (Dey and Pridham, 1972). Therefore, although α-galactosidase was active, when water stress was imposed after the start of galactomannan hydrolysis, the presence of galactose in the endosperm would inhibit its action.

Mobilisation of embryo reserves

The mobilisation of the endosperm and embryo reserves follows a time-dependent pattern that correlates the metabolic events with one another and with the completion of germination (Leung *et al.*, 1981). Galactomannan mobilisation started upon radicle protrusion, after about 25 h from seed imbibition. Before the start of galactomannan hydrolysis there is a slight decline in the embryo free sugars (Reid, 1971). Before germination there is no starch in the fenugreek embryo, but during galactomannan mobilisation there is a large increase of transient starch in both cotyledons and axes (Reid, 1971; Bewley *et al.*, 1993). At later times during seed development the embryo starch is remobilised through the action of α-amylase, which has been identified as a single band on IEF of pI 5.1 (Bewley *et al.*, 1993).

The endosperm galactomannan mobilisation was followed by the mobilisation of embryo reserves, proteins, lipids and phytate (Leung *et al.*, 1981). Galactomannan hydrolysis was followed by the deposition of starch in the embryo. The hydrolysis of cotyledon proteins started after about 30 h from imbibition. At the same time amino acids accumulated in the embryo axis, while in cotyledons the accumulation of amino acids took place later, suggesting an initial rapid uptake of the amino acids by the axis. The phytate started declining in cotyledons 50 h from imbibition and at the same time there was a slight decline in the axis apparently through the action of phytase, the activity of which started increasing after 40 h of imbibition. This metabolic event was followed by lipid hydrolysis. Lipid content, the majority of which is located in cotyledons, is approximately 8 per cent of the seed's dry weight. Concomitant with lipid hydrolysis was the increased activity of the isocitrate lyase.

Recently, the activities of α-galactosidase and endo-β-mannanase (Giammakis and Spyropoulos, unpublished data) have been detected in the fenugreek embryo. The activity of α-galactosidase was very low and did not change much during the course of the embryo growth. In contrast, endo-β-mannanase activity increased with imbibition time in both cotyledons and axes.

Tissue cultures

Fenugreek tissue and cell cultures have been used for either plant regeneration or for the production of secondary products of economic interest. Among these products are diosgenin and trigonelline: a saponin and an alkaloid with therapeutic properties, which are constituents of fenugreek seeds (Cerdon *et al.*, 1996; Merkli *et al.*, 1997; Oncina *et al.*, 2000).

The development of fenugreek calli has been achieved after shoot or root culture from 4-day-old seedlings upon culturing on Gamborg's B-5 modified medium supplemented with hormones. From these calli have been produced cell suspension cultures, the content of which in trigonelline was appreciably higher than that of the calli (Radwan and Kokate, 1980). Also, for diosgenin production hair root cultures (Merkli *et al.*, 1997) and cultures from calli, which were developed from leaves, stems and roots isolated from 30-day-old seedlings, have been established with *Agrobacterium rhizogenes* strain A4 (Oncina *et al.*, 2000).

Apart from the production of trigonelline, tissue cultures have been used for *T. corniculata* L. (Piring) and *T. foenum-graecum* L. (Methi) regeneration. In this case, calli were produced using leaves as explants. The explants were grown on Murashige and Skoog medium supplemented with casein hydrolysate or coconut milk. The first resulted in an increased number of differentiated organs per callus (Sen and Gupta, 1979).

Regeneration of shoots have also been achieved from fenugreek protoplasts (Xu *et al.*, 1982). Protoplasts were isolated from the root apices of 48-h-imbibed seeds. The first divisions of root fenugreek protoplasts were observed after a 3–4 day culture and subsequent divisions gave cell colonies. However, a culture of these colonies gave only roots.

References

Bewley, J.D., Leung, D.W.M., MacIsaak, S., Reid, J.S.G. and Xu, N. (1993) Transient starch accumulation in the cotyledons of fenugreek seeds during galactomannan mobilization from the endosperm. *Plant Physiol. Biochem.*, **31**, 483–90.

Campbell, J. McA. and Reid, J.S.G. (1982) Galactomannan formation and guanosine 5'-diphosphate-mannose: galactomannan mannosyltransferase in developing seeds of fenugreek (*Trigonella foenum-graecum* L., Leguminosae). *Planta*, **155**, 105–11.

Cerdon, C., Rahier, A., Taton, M. and Sauvaire, Y. (1996) Effect of tridemorph and fenpropimorth on sterol composition in fenugreek. *Phytochemistry*, **41**, 423–31.

Dey, P.M. and Pridham, J.B. (1972). Biochemistry of α-galactosidases. *Adv. Enzymol.*, **36**, 91–130.

Edwards, M., Dea, I.C.M., Bulpin, P.V. and Reid, J.S.G. (1989) Biosynthesis of legume-seed galactomannans *in vitro*. Cooperative interactions of α-guanosine 5'-diphosphate mannose-linked (1→4)-β-D-mannosyltransferase and a uridine 5'-diphosphate-galactose-linked α-D-galactopyranosyltransferase in particulate enzyme preparations from developing endosperms of fenugreek (*Trigonella foenum-graecum* L.) and guar (*Cyamopsis tetragonoloba* [L.] Taub.). *Planta*, **178**, 41–51.

Edwards, M., Scott, C., Gidley, M.J. and Reid, J.S.G. (1992) Control of mannose/galactose ratio during galactomannan formation in developing legume seeds. *Planta*, **187**, 67–74.

Kontos, F. and Spyropoulos, C.G. (1995) Production and secretion of α-galactosidase and endo-β-mannanase by carob (*Ceratonia siliqua* L.) endosperm protoplasts. *J. Exp. Bot.*, **46**, 577–83.

Kontos, F., Spyropoulos, C.G., Griffen, A. and Bewley, J.D. (1996) Factors affecting endo-β-mannanase activity in the endosperms of fenugreek and carob seeds. *Seed Sci. Res.*, **6**, 23–9.

Leung, D.W.M., Bewley, J.D. and Reid, J.S.G. (1981) Mobilization of the major stored reserves in the embryo of fenugreek (*Trigonella foenum-graecum* L., Leguminosae), and correlated enzyme activities. *Planta*, **153**, 95–100.

Malek, L. and Bewley, J.D. (1991) Endo-β-mannanase activity and reserve mobilization in excised endosperms of fenugreek is affected by volume of incubation and abscisic acid. *Seed Sci. Res.*, **1**, 45–9.

Meier, H. and Reid, J.S.G. (1982) Reserve polysaccharides other than starch in higher plants. In F.A. Loewus and W. Tanner (eds), *Encyclopedia of Plant Physiology (new Series)* 13A, Springer-Verlag, pp. 418–71.

Meier, H. and Reid, J.S.G. (1977) Morphological aspects of galactomannan formation in the endosperm of *Trigonella foenum-graecum* L. (Leguminosae). *Planta*, **133**, 243–8.

Merkli, A., Christen, P. and Kapetanidis, I. (1997) Production of diosgenin by hairy root cultures of *Trigonella foenum-graecum* L. *Plant Cell Rep.*, **16**, 632–6.

Oncina, C., Botía, J.A., Del Río, A. and Ortuño, A. (2000) Bioproduction of diosgenin in callus cultures of *Trigonella foenum-graecum* L. *Food Chem.*, **70**, 489–92.

Radwan, S.S. and Kokate, C.K. (1980) Production of higher levels of trigonelline by cell cultures of *Trigonella foenum-graecum* than by the differentiated plant. *Planta*, **147**, 340–4.

Reid, J.S. (1971) Reserve carbohydrate metabolism in germinating seeds of *Trigonella foenum-graecum* L. (Leguminosae). *Planta*, **106**, 131–42.

Reid, J.S.G. and Bewley, J.D. (1979) A dual role for the endosperm and its galactomannan reserves in the germinative physiology of fenugreek (*Trigonella foenum-graecum* L.), an endospermic leguminous seed. *Planta*, **147**, 145–50.

Reid, J.S.G., Davies, C. and Meier, H. (1977) Endo-β-mannanase, the leguminous aleurone layer and the storage galactomannan in germinating seeds of fenugreek *Trigonella foenum-graecum* L. *Planta*, **133**, 219–22.

Reid, J.S.G., Edswards, M.E., Gidley, M.J. and Clark, A.H. (1992) Mechanism and regulation of galactomannan biosynthesis in developing leguminous seeds. *Biochem. Soc. T.*, **20**, 23–6.

Reid, J.S.G., Edswards, M.E., Gidley, M.J. and Clark, A.H. (1995) Enzyme specificity in galactomannan biosynthesis. *Planta*, **185**, 489–95.

Reid, J.S.G. and Meier, H. (1970). Chemotaxonomic aspects of the reserve galactomannan in leguminous seeds. *Z. Pflanzenphysiol.*, **62**, 89–92.

Reid, J.S.G. and Meier, H. (1972) The function of the aleurone layer during galactomannan mobilisation in germinating seeds of fenugreek (*Trigonella foenum-graecum* L.), crimpson clover (*Trifolium incarnatum* L.) and lucerne (*Medicago sativa* L.). A correlative biochemical and ultrastructural study. *Planta*, 106, 44–60.

Reid, J.S.G. and Meier, H. (1973) Enzymic activities and and galactomannan mobilisation in germinating seeds of fenugreek (*Trigonella foenum-graecum* L.). *Planta*, 112, 301–8.

Scherbukhin, V. D. and Anulov, O.V. (1999) Legume seed galactomannans. *Applied Biochem. Microbiol.*, 35, 257–74.

Spyropoulos, C.G. and Reid, J.S.G. (1985) Regulation of α-galactosidase activity and the hydrolysis of galactomannan in the endosperm of fenugreek (*Trigonella foenum-graecum* L.) seed. *Planta*, 166, 271–5.

Spyropoulos, C.G. and Reid, J.S.G. (1988) Water stress and galactomannan breakdown in germinated fenugreek seeds. Stress affects the production and the activities *in vivo* of galactomannan hydrolysing enzymes. *Planta*, 174, 473–8.

Zambou, K. and Spyropoulos, C.G. (1989) D-Mannose uptake by fenugreek cotyledons. *Planta*, 179, 473–8.

Zambou, K. and Spyropoulos, C.G. (1990) D-galactose uptake by fenugreek cotyledons. Effect of water stress. *Plant Physiol.*, 93, 1417–21.

Zambou, K. and Spyropoulos, C.G. (1993) Saponin-like substances inhibit α-galactosidase production in the endosperm of fenugreek seeds. A possible regulatory role in endosperm galactomannan degradation. *Planta*, 189, 207–12.

4 Cultivation

Georgios A. Petropoulos

Climate and soil

Climate

Although fenugreek is a native of the Mediterranean region of Europe, it extends to central Asia and North Africa as well. It is also grown very satisfactorily in central Europe, UK and USA. This wide distribution of its cultivation in the world is characteristic of its adaptation to variable climatic conditions and growing environments. Fenugreek is suitable for areas with moderate or low rainfall. A temperate and cool growing season without extreme temperatures is favourable for the best development of fenugreek. It can tolerate 10–15°C of frost (Duke, 1986).

Fenugreek is fairly drought resistant (Talelis, 1967) and fairly frost sensitive (Talelis, 1967; Bunting, 1972). According to Del' Gaudio (1952), fenugreek adapts well during summer droughts and the wet and raining winter, while it does not like severe winter and raining summer, but it is resistant to winter cold, especially when it is covered with snow. Talelis (1967) reports that in Greece fenugreek is generally grown as a winter crop in areas with mild winter and as a spring crop in areas with soil that keeps moisture in the summer. Duke (1986) reports that fenugreek in areas with mild winters is best sown in fall to mature in spring. Also, Rouk and Mangesha (1963) notice that in Ethiopia fenugreek is grown primarily in regions where the climatic conditions approach those of the Mediterranean area. The climate of these regions is mostly subtropical and is characterised by a wet followed by a dry season. Also, they report that the annual rainfall in the areas where fenugreek is grown is in the range 10–60 in. further Allen and Allen (1981) have noticed a range of 20–60 in., while the area of widest distribution seems to fall within the 20–40 in. in the rainfall belt. Perkins (1962) reports that in India fenugreek is normally grown as a winter annual in areas described as tropical savannah and humid subtropical, with the following temperature conditions:

- hot summer and cool winters
- hot summer and mild winters.

Sinskaya (1961) reports that in Transcaucasia fenugreek reaches mountain altitudes of up to 1,300–1,400 m and in Ethiopia 3,000 m, but its main zone of distribution in that country is between 2,150 and 2,400 m. Duke (1986) reports that fenugreek, ranging from cool temperate steppe to wet through tropical very dry forest life zone, is reported to tolerate an annual precipitation of 3.8–15.3 dm and an annual mean temperature of 7.8–27.5°C. We cultivated fenugreek successfully in England in an area with an annual rainfall of around 700 mm and an average temperature for the growing season from 7–16°C (minimum 1.5 and maximum 24°C and an altitude of 80–175 m) (Petropoulos, 1973). There are indications of the possible benefit of colder nights on the sapogenin content of the seed (Fazli and Hardman, 1968).

The conclusion is that fenugreek evolved in areas that have a pronounced temperate climate with mild winters and cool summers. However in the growing areas, warm and dry conditions are desirable for field ripening of the pod.

Soil

Fenugreek does not require specific soil conditions, however one of the most important characters of a good fenugreek soil is its capacity to supply sufficient moisture throughout the growing season.

Rosengarten (1969) states that fenugreek is grown best in well-drained loams, Piper (1947) states the same adding that it is not very exact. Duke (1986) reports also that fenugreek grows fairly well on gravely or sandy soils and it is not adaptable to heavy clay or soil that becomes hard and it is fairly tolerant to salt. Bunting (1972) reports that heavy and wet soils are unsuitable for cultivation of fenugreek and mentions as an optimum pH 8–8.5. According to the appropriate Polish Institute (Anonymous, 1987) suitable soils for a successful cultivation of fenugreek are those where alfalfa grows well, as well as rendzinas, loss, alouvian, sunny and protected from winds, while unsuitable are gold, heavy, wet, very light and dry soils. Orvedahl (1962) notices that the areas of Ethiopia, India and Turkey, where fenugreek is cultivated, are characterised by soil types that are closely related to the great soil groups of the Mediterranean area, which he describes as following:

- Mountain soils of Brown Forest, Terra Rosa, and Rendzina soil regions with Lithosols, including Podzolised and Alpine Meadow soils at high elevations.
- Reddish Prairie, Reddish Chestnut and Reddish Brown.

The conclusion is that for successful fenugreek cultivation well drained loams and generally slightly alkaline soils are ideal, lime application in some strongly acid conditions may be necessary.

Sowing

It is well known that the final result of any legume crop, like fenugreek, will be satisfactory if the supply of a reliable seed is insured and better sowing practices are followed. The failure of an individual viable seed to produce a plant may be due to seed hardness, poor seedbed preparation, sowing too deep, inadequate moisture mainly after germination, freezing, competition for light and nutrients with other fenugreek seedlings or weeds etc. So, insuring of reliable seeds, seedbed preparation, sowing techniques and postsowing management should be patterned to minimise losses from these causes.

Soil preparation

Deep plowing and thorough harrowing are essential for a successful soil preparation before a fenugreek sowing (Duke, 1986; Anonymous, 1987). An ideal seedbed is moist and fairly firm. It should be sufficiently fine and granular not powdery, for good seed coverage when compacted.

Plowing and disking are the usual practices for a good preparation. Plowing may not be necessary and can be omitted when fenugreek follows most cereal crops, because a satisfactory seedbed can be prepared rapidly and at low cost, by only disking and harrowing. Also compaction prior to sowing is not necessary, except if the soil moisture is limited.

Seeds

In the growing of a fenugreek crop the use of reliable seeds is very important, both to ensure quality and identity. The quality of the seeds depends on various characteristics, especially genuineness, purity and viability. Heeger (1989) suggests that for fenugreek seed to be suitable for sowing, it should possess at least 95 per cent seed purity and 80 per cent germination ability. Other points of more or lesser importance are size, colour, one thousand seed weight, source of the seeds etc. Especially for fenugreek the percentage of 'hard' seeds is also to be considered seriously, since late emerging seedlings are likely to perish due to competition or winter injury.

For these reasons, fenugreek seeds should be purchased under guarantee, although there is a limited supply of certified fenugreek seed of named varieties. Fenugreek seed should be inoculated with the proper *Rhizobium* bacterium (see section on 'Nodulation').

Methods of sowing

Two methods of sowing, namely broadcasting on the surface of the soil and drilling are applied in the case of fenugreek sowing. In the first method, the seed is sown over the surface of the pre-pared soil either by seed tubes that are set to broadcast the seed from a height of about 2 ft above the soil, or by a broadcast seeder where the front roller compacts the seed bed and the rear roller covers the seed and it compacts the soil. Traditionally the seed that is broadcast sown is trampled into the ground. The second method is the drill with seeder attachment modified by extending the seed tubes to within 5–10 cm of the soil surface, this is a significant advance in sowing and is recommended especially under adverse conditions.

The soil is watered immediately after sowing in both methods, if rain is not expected.

Spacing and seed rate

Uniformly distributed fenugreek plants are necessary for maximum yield.

Row planting of fenugreek has more advantages than the solid one, such as more erect plants, lighter seed rates, better penetration of chemicals, lower humidity on the plant canopy etc. Plant density within the row also influences the yield and is controlled by the seed rate.

Optimum spacing on the row and within the rows depend mainly upon soil texture, depth of sowing, fertility, available moisture, temperature and variety. High densities favour monostalk plants, while lower densities favour multistalk plants (see Figure 4.1). Rosengarten (1969) and Duke (1986) recommend that fenugreek plants be spaced in rows 45 cm apart having 8 cm within rows and a seed rate of 22.5 kg/ha for broadcast, while Talelis (1967) and Bunting (1972) suggest the sowing of fenugreek in rows 30–50 cm apart with a seed rate 40–67 kg/ha. Piper (1947) recommends a seed rate of 17–22 kg/ha for seed production and 35 kg/ha for green manure, while the appropriate Polish Institute (Anonymous, 1987) suggests rows 30–40 cm apart with a seed rate of 15–20 kg/ha. Mohamed (1990) found that the number of branches, aver-age plant weight and pod number increased whereas plant height was unaffected by increase in row width from 10 to 30 cm, while the highest seed yield (1,650 kg/ha) was achieved with a row

Figure 4.1 Multistock and monostock plants of fenugreek, due to the corresponding low and high plant density (1 = monostock, 2 = multistock).

width of 20 cm. Dachler and Pelzman (1989) suggest sowing of fenugreek in rows 25 cm apart with a seed rate of 25 kg/ha.

The above contradiction regarding the spacing and the seed rate of fenugreek and the fact that some growers have generally advanced the opinion that close spacing results to higher profit from the increased plant population, led to a trapezoidal spacing experiment being carried out (Petropoulos, 1973), as was described by Bleasdale and Nelder (1960). It was found that for fenugreek an optimum spacing of 0.0631 m^2/plant was needed. So, the optimum plant density that must be applied to obtain the maximum yield of fenugreek seed is 158,480 plants/ha. Estimating that 1 kg seeds of fenugreek contains approximately 35,000–53,000 seeds for different varieties and conditions of cultivation with 50 per cent losses for different reasons (hard seeds, reduced seed germination capacity, unreleased cotyledons etc.), the maximum seed rate of a broadcast seeding is 10–14 kg/ha. When drill seeded, lesser quantities of seed could give satisfactory results.

Time of sowing

As fenugreek is fairly drought resistant (Talelis, 1967) and fairly frost resistant (Talelis, 1967; Rosengarten, 1969), it is generally grown as a winter crop in areas with mild winter and as a spring crop in areas with soil that keeps moisture in the summer (Talelis, 1967). In India it is grown as a traditional winter crop (Pareek and Gupta, 1981) similarly in Egypt (Rizk, 1966). Under certain conditions early fall sowings are satisfactory, however spring sowings are recommended for all areas with prolonged periods below freezing, as in Germany where sowing takes place in April (Dachler and Pelzmann, 1989; Heeger, 1989). But the risk of poor germination increases and yield decreases as planting time prolongs into spring. Rathore and Manohar (1989) found that in India early sowing in the fall (15 October) gave a higher yield than late sowing (14 November).

An experiment was carried out in England (Petropoulos, 1973) to test the effect of three different dates of planting with monthly intervals starting from March in four cultivars in a spring crop of fenugreek. The conclusions drawn from this experiment are:

i   There is a linear response between earliness of sowing and earliness of flowering and consequently of maturity, as it is presented in Figure 4.2.
ii   There was insufficient time for the majority of the late sowing plants to attain full maturity, especially those of the late cultivars.
iii   Fenugreek by England conditions (Bath area) can be sown from mid-March to mid-April when soil conditions allow and risks of severe frost recede.

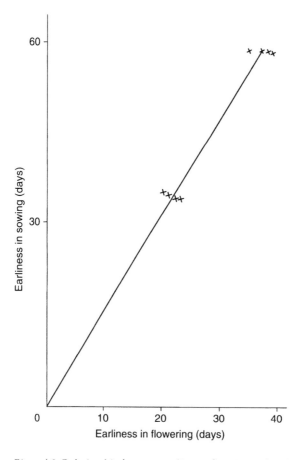

Figure 4.2 Relationship between earliness of sowing and earliness of flowering and consequently of maturity (based on sowing in mid-May).

Nodulation

The extraordinary property of legumes to fix atmospheric nitrogen (N) by symbiosis with *Rhizobium* was known to botanists and agronomists of the last century (Hallsworth, 1958).

Fenugreek is grown mostly in subtropical areas and attempts to extend its culture to new soils as a temperate crop often failed. There is a good possibility that many of those failures were directly attributable to the lack of effective nodule bacteria. So, when fenugreek is introduced into a new area, artificial inoculation is commonly applied the first year or two of planting (Anonymous, 1961).

It is well known that there are many kinds of nodule bacteria, homologous and heterologous, as the various leguminous have their preferences (Fred *et al.*, 1932; Pattison, 1972) and *Rhizobium meliloti* is homologous with *Trigonella foenum-graecum* that is able to form an effective symbiotic association with fenugreek (Subba-Rao and Sharma, 1968). This *Rhizobium* nodulates also alfalfa, sweet clover, burclover, button-clover, burrel medic and other species of *Medicago*, *Trigonella* and *Melilotus*, but no other species of Leguminosae (Burton, 1975).

Rhizobium meliloti is one of the six designated species of nodule bacteria in the family *Rhizobiaceae*. It is a typically fast growing *Rhizobium*, aerobic, nonspore forming gram-negative, motile robs with peritrichous flagellation. These Rhizobia are grown best when cultured on extracts of yeast, malt or other plant materials that provide readily available N and growth factors. Strains of *R. meliloti* are the most sensitive to acidity and grown very poorly at a pH of 5.0 or below. Its nodules are at first spherical but later branch into a two-lobed or a fan-shaped structure within 4–5 days of their initiation (Burton, 1975).

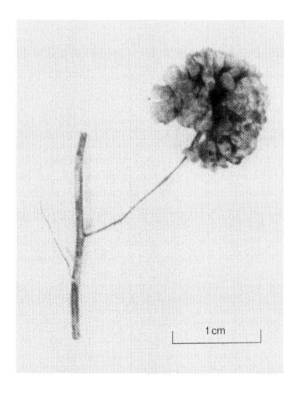

Figure 4.3 A typical nodule of *Rhizobium meliloti* 2012 on fenugreek.

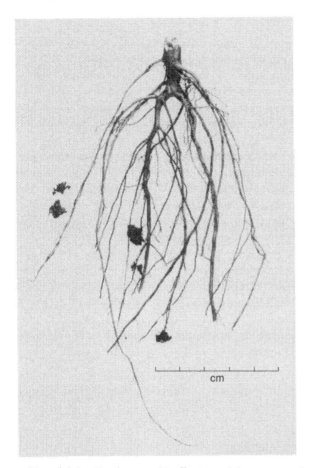

Figure 4.4 Small and scattered ineffective nodules over secondary roots of fenugreek.

The nodule is the focal point of reaction between Rhizobia and the fenugreek plant. There are effective and ineffective nodules. The first are usually large, elongated, often clustered on the primary roots (see Figure 4.3), while the second ones are usually small and are mainly scattered over the secondary roots (see Figure 4.4). Both effective and ineffective nodules frequently occur simultaneously on the plant root system. Burton (1975) claimed that a leguminous plant's susceptibility to nodulation is related to its pollination characteristics and postulated that cross-pollinating species carry genetic characters that make them promiscuous with diverse Rhizobia, whereas in self-pollinating species, like fenugreek, the characters permitting nodulation are limited or carried as recessives.

As fenugreek is cultivated in different environments, it is very likely that certain strains of Rhizobia are better adapted than others in these various conditions. So, it is necessary to find the proper strains of Rhizobia by selection or genetic manipulation for all these special conditions. Hardman and Petropoulos (1975) found that the strain *R. meliloti* 2012 obtained from the Rothamsted collection and originating from Sidney University, nodulates fenugreek satisfactorily (Pattison, 1972).

Rhizobia are applied either to fenugreek seeds or to soil. The first method is preferable as it is easy and convenient to implant the Rhizobia into the soil where the roots of the young seedlings will grow. The seeds should be covered uniformly with vigorous inoculum of Rhizobia. The viable rhizobial content of inoculants decreases rapidly with time, unless refrigerated. Effective nodulation under any particular set of conditions depends greatly upon the type of inoculum employed and the method of inoculating the fenugreek seeds. There are in commerce three types of *R. meliloti* inocula available to growers: the moist powder peat-base that is the most popular, the liquid or broth culture and an oil-dried rhizobial preparation absorbed in pulverised vermiculite (Burton, 1975).

There are three most used methods for inoculation for fenugreek and other leguminous seeds. The *sprinkle*, where the seeds are sprinkled with a small amount of water and the dry inoculant powder is mixed thoroughly with the moistened seed. The *slurry*, where the inoculum is suspended in sufficient water to cover the seed uniformly. The *waterless*, where the powdered inoculant is added directly to the seed in the drill hopper without using any water. Awasthi and Narayana (1984) found that sprays of sucrose plus boric acid enhanced inoculation and N fixation of fenugreek. Hardman and Petropoulos (1975) used a pure culture of the *Rhizobium* in skimmed milk for inoculation of the moistured seeds, which were dried away from light and heat and sown immediately. On a global basis little arguments exist that inoculation is needed in the majority of agricultural soils, as the difference between inoculated and uninoculated plants is often markedly apparent. Campbell and Reid (1982) found in Egypt that the amount of atmospheric N fixed by fenugreek was 42.4 kg/acre, which was more than double in comparison

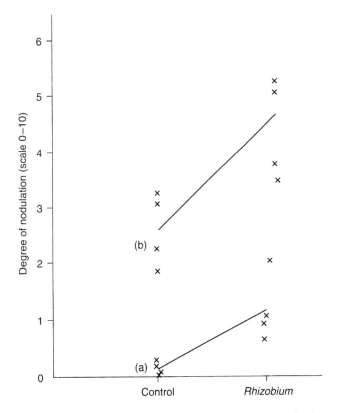

Figure 4.5 Degree of nodulation of fenugreek plants with *Rhizobium meliloti* 2012 in (a) virgin and (b) non virgin soil.

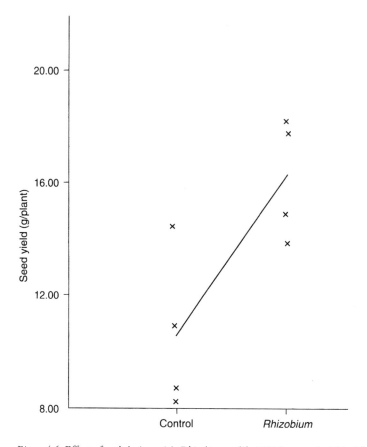

Figure 4.6 Effect of nodulation with *Rhizobium meliloti* 2012 on seed yield of fenugreek plants.

with that of soybean. The cost of inoculating fenugreek seed is low depending on the method used, the farm application costs around $1/ha.

Hardman and Petropoulos (1975) tested inoculated and uninoculated seeds of the four cultivars in virgin and non virgin soil and the conclusion drawn from this experiment was that the inoculated fenugreek plants were taller and well nodulated, especially in the case of non virgin soil (Figure 4.5), with a higher seed yield (Figure 4.6), but delayed in maturing. The seed of inoculated plants had a higher crude protein content and in agreement with this a higher germ/husk index and a lower mucilage content than seed from the uninoculated controls and there was no indication of any interaction between the tested cultivars and *R. meliloti*.

The final conclusion is that the inoculation of the fenugreek seed, before sowing, help the insurance of N fixation, especially when fenugreek has not been grown in the area previously. However the effectiveness of nodulation is generally improved with additional Rhizobia.

Depth of sowing

Soil moisture, soil type and time of planting influence the optimum depth of fenugreek sowing for total emergence. Planting too deep was frequently the cause of sowing failures in fenugreek cultivation. The appropriate Polish Institute (Anonymous, 1987) suggests a sowing depth of

1.0–1.5 cm, while Dachler and Pelzmann (1989) report a depth of maximum 2 cm. Deeper seed plantings are recommended for coarse textured soils, subject to drought or in arid areas. Shallow depths are best when moisture conditions are favourable as in spring and greater depths are recommended when moisture conditions are less favourable. For shallow sowing depths and low moisture conditions compaction may be necessary with a corrugated roller. Also, in shallow depths there was a higher proportion of unreleased cotyledons from the husk, where the majority of these seedlings usually die.

Seed germination and the first growth

According to our observations the curve protrusion of the radicle for more than 5 mm is considered as sign of fenugreek seed germination, because only if the radicle has developed to such a length can the cotyledons may be counted upon to follow. Zade *et al.* (1990) suggest a new germination testing procedure. In order that fenugreek seeds may germinate perfectly, it is necessary that they are well developed and have vigorous germs and an abundant supply of stored food.

Three conditions are necessary for germination: (i) sufficient moisture, (ii) sufficient oxygen and (iii) sufficient heat. An interrelationship exists among them. The presence of available water is absolutely necessary for fenugreek seed germination. The minimum amount of this water for four cultivars of fenugreek is presented in Table 4.1.

In the soil fenugreek seeds follow the epigeal way of germination. So, after the absorption of water and swelling of the starch-free and high thickened cells of endosperm, the radicle is the first part of the embryo to elongate and emerge from the husk and enters the soil, becoming the primary root and developing secondary roots. The cotyledons are pulled above the soil by the elongation of the hypocotyl, which makes a crock (curve), while the epicotyl is characteristically absent in the first stage of growth of fenugreek seedlings. The husk usually releases the cotyledons into the soil, but sometimes the cotyledons are not detached and the husk emerges covering the cotyledons. In this case, if it is not raining, irrigation may be necessary, as, according to our observations, the average of unreleased cotyledons reach approximately 20 per cent and the majority of these seedings usually die (Petropoulos, 1973). The cotyledons in fenugreek plants serve as foliage leaves and in certain cases remain for the whole life of the plant.

The time of germination in soil usually varies from 3–10 days. Dachler and Pelzmann (1989) report that fenugreek seeds germinate 10 days after sowing, while the appropriate Polish

Table 4.1 Determination of the water requirements for seed germination among four breeding cultivars of fenugreek

No.	Cultivar	One hundred seed weight (g)	Absorbed water before germination starts	
			Percentage of seed weight	Per 100 seeds (g)
1	Fluorescent	2.9	148	4.3
2	Ethiopian	2.6	155	4.0
3	Moroccan	2.7	160	4.3
4	Kenyan	1.7	176	3.0

Table 4.2 Determination of the hardness of fenugreek seeds due to the drying conditions

Seed sample (seeds in pod immediately after harvesting)	Characteristics	Drying conditions					
		Room temperature for 20 days (control)		35°C for 48 h		50°C for 24 h	
		Final moisture content	Hard seeds (%)	Final moisture content	Hard seeds (%)	Final moisture content	Hard seeds (%)
RH 3142	Kenyan cultivar, origin pods of plants produced from hard seeds of RH 2926	10.1	8	7.5	34	5.3	72
RH 3143	Kenyan cultivar, origin pods of plants produced from soft seeds of RH 2926	10.2	6	7.5	32	5.3	68

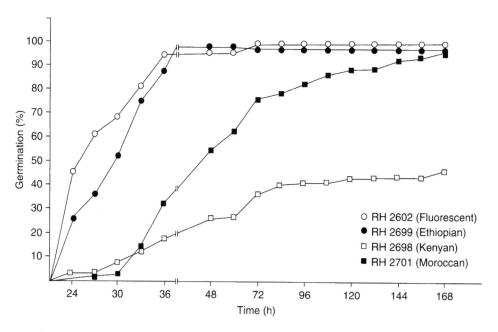

Figure 4.7 Prolonged period of seed germination of Moroccan and especially of Kenyan cultivar of fenugreek, due to their higher percentage of hard seeds.

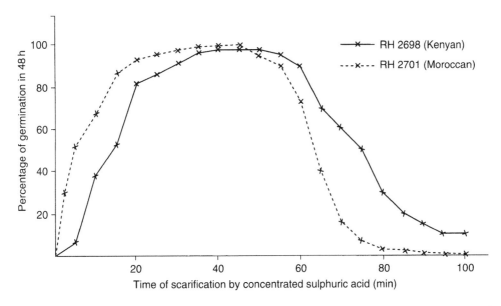

Figure 4.8 Relationship between scarification time by concentrated sulphuric acid and percentage of fenugreek seed germination, with optimum time in 35–40 min.

Institute (Anonymous, 1987) reports 10–14 days, although this also depends mainly on the soil conditions (temperature, available moisture etc.), the osmotic concentration of the media surrounding the seeds, the depth of sowing (earlier in shallow sowing), the quality of the seed (germination energy) and the variety of fenugreek (seed coat, micropyle etc.). But there is a percentage of named 'hard' fenugreek seeds that are naturally slow to germination, because they are unable to absorb water rapidly. In fact they start an irregular prolonged germination period of even more than six months. This phenomenon is characteristic of the variety, but it also depends upon external factors like the artificial drying of the pods, where we noticed that the faster the rate of drying the greater the proportion of hard seeds (see Table 4.2).

Among the four tested cultivars named Fluorescent, Ethiopian, Kenyan and Moroccan, the Moroccan cultivar and mainly the Kenyan one possess the highest percentage of hard seeds, as is presented in Figure 4.7, by a prolonged germination period. To ensure an increase and acceleration of the hard fenugreek seed germination we found that a scarification with concentrated sulphuric acid for 35–40 min gives the best results, as is presented in Figure 4.8. It was found that if the proportion of hard seeds exceeds 40 per cent, the fenugreek seed should be scarified before planting (Petropoulos, 1973). Six to ten days after the fenugreek germination the seedlings produce the first leaf, which is usually simple, there is still no noticeable epicotyl as the first trifoliate leaf is formed after a further 5–8 days (see Figure 4.9).

Plant growth

After the seed germination and the first growth of the seedling, follows the main plant growth, which includes the development of stems, flowers, pods and seeds. The fenugreek has an

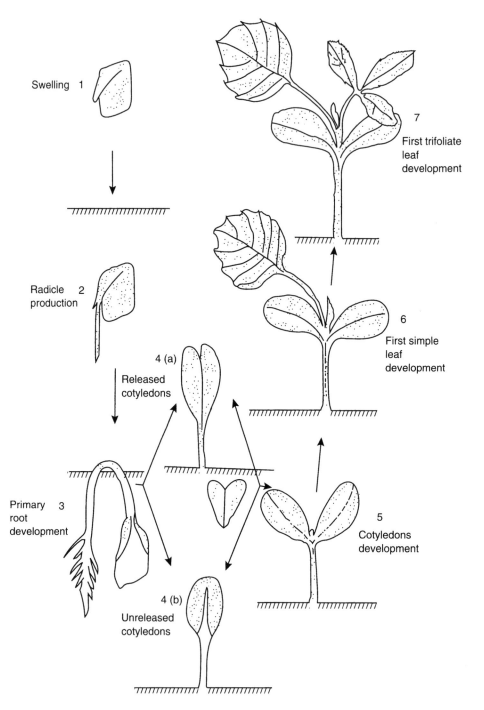

Swelling 1

Radicle 2
production

Released
cotyledons

4 (a)

Primary 3
root
development

4 (b)

Unreleased
cotyledons

5

Cotyledons
development

6

First simple
leaf
development

7

First trifoliate
leaf
development

Figure 4.9 The first growth habit of a fenugreek seedling.

indeterminate growth habit, which means the growth continues from the terminal and axially buds, while the flowering and formation of pods are both in progress.

Stems

The stems of fenugreek are erect, hollow, with dark anthocyanin or complete green. The stems according to variety, soil fertility and plant density are either monostalk without secondary shoots, or multistalk where many shoots arise from the basal and higher nodes. In some cases the main shoot does not differ markedly from the secondary shoots. This last plant shape is resistant to lodging and produces an increased number of pods/plant.

Flowers

The flowering of fenugreek, according to variety, climate and season of sowing starts approximately 35–40 days from the sowing. The flowers of fenugreek are seated in the leaf axils mostly paired (twin), more rarely solitary. There are two kinds of flower shoots. The common ones with axillary flowers only that follow the indeterminate growth habit, where the shoot apex

Figure 4.10 'Blind' shoot of fenugreek with axillary and terminal flower.

continues to differentiate both vegetative and floral organs and the 'blind' shoots with axillary and terminal flower bud, which become tip bearers (see Figure 4.10). Each flower consists of a calyx, a corolla, ten stamens and a pistil. The calyx tube consists of five undivided sepals, ending with five teeth about as long as the tube. The corolla is highly evolved and consists of five petals: a large standard (banner), two lateral wing petals and two fused petals that form the keel (see Figure 4.11). A sectional view of the fenugreek flower showing the relative position of the stamens and pistil appears in Figure 4.12.

There are also two kinds of fenugreek flowers:

1   *Cleistogamous (closed) flowers.* In this category belong the majority of fenugreek flowers (see Figure 4.13) in which the keel remains closed during the entire life of the flower, while the standard and wings open some hours per day. These flowers, described in Chapter 5, are usually self-pollinated.
2   *Aneictogamous (open) flowers.* These are flowers in which all the parts of the corolla remain open almost continuously. These flowers usually present some abnormalities, that is, the corolla fused on the calyx or two pistils etc., they are less than one per cent of the total number of fenugreek flowers and are usually born on the 'blind' shoots (see Figure 4.14) and offer many opportunities for cross-pollination.

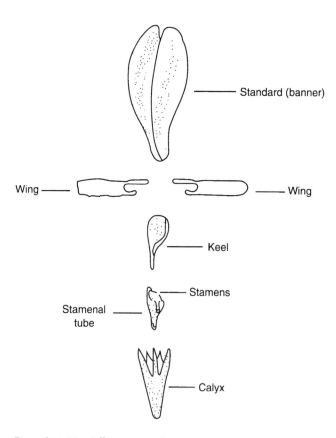

Figure 4.11 The different parts of the corolla of a fenugreek flower.

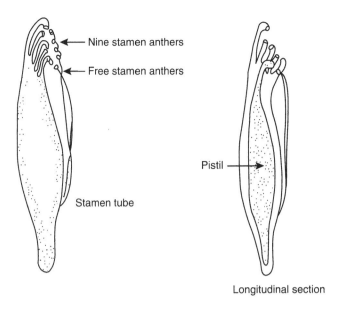

Longitudinal section

Figure 4.12 The relative position of the stamens and pistil of a fenugreek flower.

Figure 4.13 A 'cleistogamous' (closed) flower of fenugreek, that favours self-pollination.

Figure 4.14 An 'aneictogamous' (open) flower of fenugreek, that favours cross-pollination.

In the cleistogamous flowers of fenugreek there are four distinguished stages of development:

1   *First stage (flower bud)*. This stage starts from the appearance of the flower bud until the petals reach the length of the calyx teeth. During this stage the anthers are closed and arranged in two circles (upper and lower), while both are lower than the stigma of the pistil. Each circle is composed of the anthers of five alternate stamens. The stigma in this stage is on the begging to be receptive, so it is suitable for emasculation in order to avoid completely selfing or undesirable crossing in the case of artificial pollination, especially for critical breeding studies. The duration of this stage is 3–4 days.

2   *Second stage (main development)*. All the flower parts of fenugreek show a vigorous development during this stage. The corolla increases in length but remains straight and its colour is yellow. Some openings of the standard and the wings are noticed at the end of this stage for some hours daily, and this time is the main opportunity for cross-pollination for this type of flower. The stamens elongate quickly and form the staminal tube, their anthers reach and exceed the stigma, while they start to rupture and lightly dust the stigma with pollen. The pistil is also developed but slowly in comparison with the stamens and its stigma finally remains in a lower position than the anthers. This is the main stage of development and its duration is 2–3 days.

3   *Third stage (pollination)*. The corolla nearly takes its final curved shape and size and daily openings of the standard and wings are noticed, but the corolla's colour still remains yellow. The rupture of the anthers is continued and completed and they dust the stigma with their pollen. The stigma is completely receptive. This is chiefly the pollination stage and its duration is 2–3 days.

4   *Fourth stage (fertilisation)*. The corolla takes its final size and its colour may turn to white. The opening of the standard and wings for some hours daily is also noticed. The anthers have completely ruptured and there is a mass of pollen on the stigma. The process of fertilisation starts. So, this is chiefly the fertilisation stage and its duration is 4–5 days.

A diagram of these four stages appears in Figure 4.15.

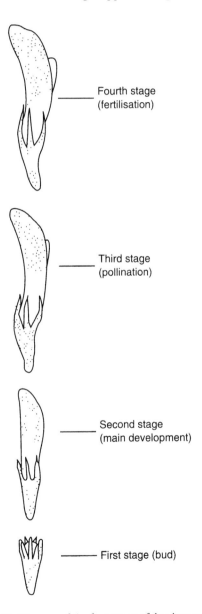

Fourth stage
(fertilisation)

Third stage
(pollination)

Second stage
(main development)

First stage (bud)

Figure 4.15 Diagram of the four stages of development in a fenugreek flower.

Pods

The pod of the fenugreek is long and pointed and has a length (excluding beak) of 60–110 mm (Ivimey-Cook, 1968) or 75–150 mm (Duke, 1986) and a width of 4–6 mm, erect or patent, linear, somewhat curved, glabrous or glabrescent with longitudinal veins. The beak persistent has a length 10–30 mm (Ivimey-Cook, 1968) or 2–3 mm (Duke, 1986).

Fenugreek plants may be divided into two classes for the number of pods per node near the top of the stem, namely 'solidary pods' when there is only one pod per node and 'twin pods' when two pods project in opposite direction from the same node of the stem (see Figure 4.16). It must be emphasised that the position of growth of these twin pods should be near the top of the stem, as on the base of the stem almost all the nodes of most varieties of the fenugreek plant possess double pods. The twin pods as it is described in Chapter 5 is a very good index of selection for higher diosgenin seed content.

For the purpose of this edition, pods up to 5.5 mm in width will be termed 'wide' and pods less than 5.5 mm will be termed 'narrow'.

The stages of pod development are described in the section of 'Harvesting'.

Seeds

Fenugreek seeds according to Wallis (1960) and Fazli and Hardman (1968) are about 2.5–6 mm long, 2–4 mm wide and 2 mm thick. They are hard, yellowish-brown, irregularly rhomboidal, round or square in outline (Fazli and Hardman, 1968), flattened and some of them fluorescent under UV light (Petropoulos, 1973).

Figure 4.16 Twin pods on the top of the fenugreek mutant plant RH 3112.

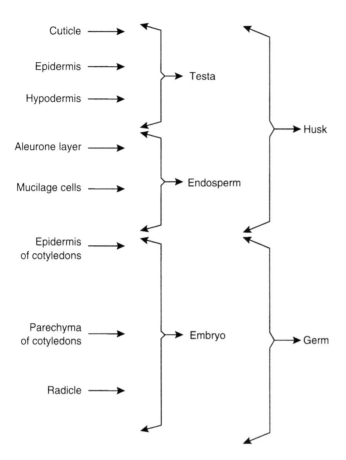

Figure 4.17 The different parts of a fenugreek seed.

Nearly in the centre of one of the long narrow sides, there is a small depression in which both hilum and micropyle are situated. This depression is continued in the form of a furrow running diagonally across parts of each of the adjoining sides, thus dividing the radicle-pocked from the remainder of the seed, in which are placed face to face the two large cotyledons, the radicle being accumbent. The embryo is yellowish and the cotyledons are surrounded by scanty, horny, dark translucent endosperm. The endosperm swells up in water to a thick gelatinous sac (Fazli and Hardman, 1968).

According to descriptions of Parry (1943), Fazli and Hardman (1968) and Reid and Bewley (1979), the different parts of the fenugreek seed are presented in Figure 4.17.

For the purposes of this edition the following terms have been accepted for fenugreek seeds:

1   'Large' when the one thousand seed weight is more than 20 g, and 'small' when this weight is less than 20 g.
2   'Rectangular' when the outline shape of seeds is approaching rectangular, and 'round' when the outline shape is approaching that of type B presented by Fazli and Hardman (1968) (see also Figure 4.18).

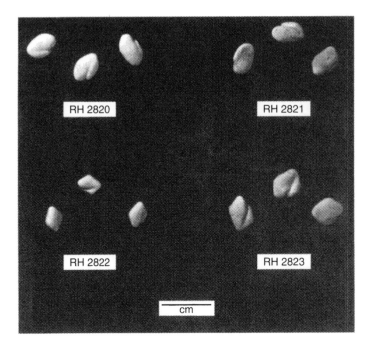

Figure 4.18 Rectangular (down) and round (upper) shape of fenugreek seeds.

3   The g/h index expresses the two decimal points of the division germ weight/husk weight, as in all the cases of a mature fenugreek seed the germ weight is higher than the husk weight.

Fenugreek seeds, especially if powdered, possess a spicy, strong and characteristic odour and their taste is slightly bitter, oily and farinaceous (Fazli and Hardman, 1968). Max (1992) emphasises their pungent aromatic properties.

Cultural practices

The cultural practices or cares of a fenugreek crop mainly include irrigation, fertilisation, weed control and disease-pest control, but only irrigation will be described in this section, as each of the remaining three practices constitute separate chapters of this volume.

Irrigation

It is well known that the highest yields are obtained when irrigation practices prevent severe plant stress and promote smooth and continuous growth process through the entire active growing period of any crop, including fenugreek. The problem is the availability of irrigation water and if it is beneficial to be used for fenugreek or for other more profitable crops.

One of the most important characters of good fenugreek soil is its capacity to supply sufficient moisture throughout the season for the active growth of the crop. Although fenugreek is fairly drought resistant (Talelis, 1967; Duke, 1986), however if rainfall plus the residual water does not cover the water requirements of a fenugreek crop, then the addition of water by irrigation is necessary and this is significant for arid and semi-arid areas. Del' Gaudio (1952) supports that if the rainfall from September–April is less than 400 mm, irrigation of fenugreek is necessary.

Irrigation requirements

Irrigation requirements for fenugreek seed or forage production are dependent upon soil depth and texture, evaporation, temperature and cropping practices. Shallow and sandy soils need more often but fewer amounts of irrigation water than the compact deep soils. Higher degrees of evaporation and temperature are required for more amounts of irrigation water. Cropping practices include plant density, and the water requirements increase on increasing this density. For this reason when fenugreek is grown as an irrigated crop, the sowing is broadcast rather thickly onto beds (Duke, 1986).

When and how much to irrigate

Irrigation must start immediately after sowing to help in seed germination and be continued when necessary. This early watering is necessary even for non irrigated fenugreek crops, if rain is not expected after sowing (Fazli and Hardman, 1968; Duke, 1986). Water supply should be at a depth that is within reach of the roots. As fenugreek possesses a shallow root system, heavy watering is not needed.

The determination of soil moisture and the inspection of plant appearance, preferably, in the morning are going to help the grower to decide a suitable time to apply irrigation. It is estimated that a water quantity of 200 m^3/ha every time for sandy soils, and 250 m^3/ha for heavier soils replicated every fortnight is sufficient for a successful fenugreek crop. Pareek and Gupta (1981) report the application of irrigation five times for the whole growing period of a fenugreek crop under Indian conditions.

Quality of irrigation water

Although fenugreek is tolerant to salt (Duke, 1986) recently Yadar *et al.* (1996) reported that irrigation with sodic water (EC 1.93 ds/m and residual sodium carbonate 12.0 me/l) resulted in a greater percentage of deduction in seed yield of fenugreek than in the more tolerant spices (fennel and black mustard).

Method of application

As far as the method of irrigation is concerned for fenugreek, both flood and spraying are usually applied (Saleh, 1996).

Varieties

General

Although the main area cultivated with fenugreek is concentrated in some countries of Asia and Africa, however it has been distributed in many countries throughout the world under different environments. So, for a successful cultivation of fenugreek, varieties that are high yielding with wide adaptability are needed.

It is well known that for self-pollinated plants, like fenugreek, distinct and uniform varieties exist. Information on the breeding work that has been performed on fenugreek for the creation of improved varieties is scanty. Reports on this aspect are few and scattered even in India where recently Edison (1995) reported that the research for spices in his country is still in its infancy. He realises the lack of advanced breeding methods and adequate genetic variability for evolving high yielding varieties, with greater stability. However, due to persistent efforts for releasing improved varieties various research institutions of India have released five fenugreek varieties in the last eight years and more have been recommended for wide cultivation (Edison, 1995). But mostly improved fenugreek varieties in India have been created and evaluated locally, the introduction of germplasm from around the world to increase genetic variability and to adapt the crop to a wide array of growing environments, has not been realised. For this reason serious efforts have been made in India to promote the import/exchange of valuable germplasm as well as varieties mainly from the Mediterranean region, in order to overcome the yield barrier (Edison, 1995).

Attempts have also been made for relevant research work on fenugreek, usually covered under the framework of massive agricultural research in different institutes and universities throughout the world. Outside India some improved varieties and cultivars have been created (Del Gaudio, 1953; Bunting, 1972; Petropoulos, 1973; Hardman, 1980; Cornish *et al.*, 1983; Saleh, 1996).

Until about 1970 the varieties of fenugreek used were directed mainly for flavouring purposes in food and as a spice. The potential industrial use of fenugreek as a source of steroidal diosgenin strengthens its position as a chemurgic crop and establishes a pattern for the development of new improved varieties.

Varieties and cultivars[1]

The varieties or cultivars of fenugreek that are used most are listed in Table 4.3.

Five varieties in Table 4.3 named as 'Co-1', 'Rajendra kanti', 'RMt-1', 'Lam Sel 1' and 'Pusa Early Bunching' (HM-57) are reported by Edison (1995) as the most interesting varieties of fenugreek in India, most of them have been recommended for a wide cultivation by farmers.

The varieties reported in Table 4.3 by Kamal *et al.* (1987) collected from different geographical regions of South India and tested spectrophotometrically for their diosgenin content of seed, ranged from 750 mg % in UM-112 to 70 mg % in UM-17. Among those with high diosgenin content are also Co 1 (650 mg %) and CVT UM TC 2336 (455 mg %), while those with low diosgenin content include UM-18 (87 mg %) and UM-75 (125 mg %).

The twenty varieties in Table 4.3 reported by Prasad and Hiremath (1985) were screened for their resistance against *Rhizoctonia solani*, and only TG-18 and UM-20 showed some tolerance, while none showed complete resistance.

The Egyptian variety 'Gharbin-6' is an old and productive one, it is the creation of the Giza Cairo Experimental Plant Station.

Del' Gaudio (1953) in Italy selected a new variety 'Ali corte' from the basic variety 'Ali lunghe' with short wings to the flower. It is more productive of fresh forage and seed.

Vaitsis (1985) in Greece evolved the variety 'Ionia' with long stems, resistant to the fungus *Sclerotinia sclerotiorum*, with high precocity, good adaptability, tolerant to cold and high yielding. It has been listed in the official *Journal of the European Communities* (Anonymous, 1996).

---

1 We use the term 'cultivar' only for genetic materials of fenugreek that have not been released yet for a wide cultivation by farmers, while the term 'variety' is used for the genetic materials that have been released for a wide cultivation usually by certified seed and have been registered in relevant catalogues.

Table 4.3 List of the most used varieties or cultivars of fenugreek in the world

No.	Varieties or cultivars	References	Country	Remarks
1	CO-1, Rajendra Kanti, RMt-1, LamSel 1, Pusa Early Bunching	Edison, 1955	India	Details in the text
2	UM-9, UM-17, UM-18, UM-23, UM-25, UM-26, UM-27, UM-32, UM-33, UM-36, UM-50, UM-52, UM-58, UM-67, UM-70, UM-75, UM-77, UM-79, UM-83, UM-84, UM-105, UM-112, UM-113, UM-114, UM-115, CVT UM-5, CVT UM-17, CVT UM-32, CVT UM-34, CVT UM-35, CVT UM TC 2336, CVT TG 1084, CVT GF 1, CVT CC, CVT NLM, NLM, CO 1, Local check, CT Lam Sel 1	Kamal *et al.*, 1987	India	Details in the text
3	RG-07, TG-3, TG-13, TG-18, TG-24, TG-34, UM-5, UM-6, UM-17, UM-20, UM-34, UM-35, UM-38, NI-01, MP-14, IC-99, LamSel 1, Local Bobes, Pusa Earlier, Bangalore-Local	Prasad and Hiremath, 1985	India	Details in the text
4	T-8	Paroda and Karwasra, 1975	India	Highly unstable especially in poor environments
5	HM-46	Singh *et al.*, 1994	India	No reduction of phenol at maturity
6	IC-74	Singh and Singh, 1974	India	Mother of the mutant 'Trailing Green'
7	'Gharbin-6'	Bunting, 1972	Egypt	Old variety
8	'Ali Lunghe', 'Ali Corte'	Del' Gaudio, 1953	Italy	Details in the text
9	Ionia	Vaitsis, 1985; Anonymous, 1996	Greece	Details in the text
10	Gouta	Haefele *et al.*, 1997	France	—
11	Barbara, Margaret, Paul	Hardman, 1980; Evans, 1989	England	Details in the text
12	Fluorescent, Ethiopian, Kenyan, Moroccan	Petropoulos, 1973	England	Details in the text

The three varieties in Table 4.3 created by Hardman (1980) named 'Barbara', 'Margaret' and 'Paul', are entered in the UK National List. Their main characteristics are the following:

1   *Barbara*: Soft seeds, fluorescent under UV light, suitable for forage production, also high in protein, fixed oils and mucilage content of seed. Diosgenin: 1.2 per cent, D-value: 89, resistant to autumnal fungi attacks.
2   *Margaret*: High average of hard seeds and may need scarification. Similar to Paul variety. Medium in protein and fixed oils content. Diosgenin: 1.5 per cent, D-value: 81.
3   *Paul*: High percentage of hard seeds and may also need scarification. Low in protein and fixed oils content. Diosgenin: 1.4 per cent, D-value: 81. Resistant to frost.

Four cultivars in Table 4.3 named 'Fluorescent', 'Ethiopian', 'Kenyan' and 'Moroccan' are described in more details as follows. They are being published for the first time.

1   *Fluorescent (RH 2602)*. This cultivar was created by continuous mass selection of a spontaneous mutation from the Ethiopian population RH 2475 with criteria the wide and long pods and the uniform plants. Its main characteristics are: large and round in outline seeds that look like the shape B as described by Fazli and Hardman (1968), these are fluorescent under UV light and this property is controlled by a single recessive gene (Petropoulos, 1973). The wide and long pods contain 10–15 seeds, they change from green to a light straw colour when ripe. This cultivar belongs to a pallida type (see Chapter 5). It possesses a high proportion of 'open' flowers. It is characterized by the absence of hard seeds. It is a very tall cultivar with a high g/h index.
    The advantages of this cultivar are the simultaneous and relative high content of the four active constituents (diosgenin, protein, fixed oils and mucilage) of seeds, and its usefulness for genetic studies, as many distinguishing morphological characters are controlled by recessive genes (Petropoulos, 1973). Also it possesses a very high specific seed weight, resistance to fungi *Ascochyta* sp. and *Oidiopsis* sp. and tolerance to Bean Yellow Mosaic Virus. The susceptibility to fungus *Heterosposium* sp., mineral deficiencies, winds, premature germination of seeds in the pods, late maturity and the quick loss of its seed viability are some of its disadvantages.
2   *Ethiopian (RH 2699)*. This cultivar was created by continuous mass selection of a spontaneous mutation from the Ethiopian population RH 2278 with criteria the wide pods and the uniform plants. It belongs to the colorata type (see Chapter 5) and reddish secondary shoots arise from the base. The pods, when ripe, take a light brown colour with 9–14 round seeds, belonging to punctate olivacea according to Serpukhova's classification (Serpukhova, 1934).
    The advantages of this cultivar are the high percentage in crude protein and fixed oils and its resistance to the fungi *Ascochyta* sp. and *Oidiopsis* sp. The susceptibility to the fungus *Heterosporium* sp., the prematurity of the seeds in the pod, the late maturity and the relatively low yielding nature, are some of its disadvantages.
3   *Kenyan (RH 2698)*. This cultivar was created by continuous mass selection of the Kenyan population RH 2591 with criteria the high proportion of twin pods on the top of the stem, the resistance to mineral deficiencies and winds, the high yielding and the property of no sprouting in the pod. The main shoot does not differ markedly from the secondary shoots, which arise from the base. This cultivar belongs to the colorata type. The pods are narrow and short and turn from slight reddish before ripening to light brown when ripe, they contain 14–20 seeds, and belong to the nanofulfa type, according to Serpukhova's classification (Serpukhova, 1934). They look like the shape C as described by Fazli and Hardman (1968).

The advantages of this cultivar are: the very high diosgenin content of seeds, the high seed yielding nature, the earliness in maturity, the absence of sprouting in pod, the resistance to damp weather and to winds because of the strong stems and the secondary shoots that arise from the base, and finally its fair resistance to the fungus *Heterosporium* sp. Among its disadvantages are included the susceptibility to attacks by the fungi *Ascochyta* sp. and *Oidiopsis* sp., attacks by the Bean Yellow Mosaic Virus and to mineral deficiencies. Also the high proportion of hard seeds that impose the need of scarification before sowing, the scattering and shattering of the seeds, the low protein content and the low g/h index.

4   *Moroccan (RH 2701)*. This cultivar was created by continuous mass selection of the Moroccan population RH 2283 with criteria the uniform plants with large seeds and high proportion of twin pods.

Although it belongs to the colorata type the stems, the petiolules and the blades of the leaves are without anthocyanin. The pods are green before ripening and turning to light straw or silver when ripe, long but narrow, containing 12–16 seeds that belong to the magnofulva type according to Serpukhova's classification (Serpukhova, 1934). They look like the shape A as described by Fazli and Hardman (1968).

The advantages of this cultivar are the earliness of ripening, the resistance to winds because of the shortness of the plant, the absence of sprouting and shedding of the seeds and the fair resistance to the fungus *Heterosporium* sp. and to mineral deficiencies. The susceptibility to fungi *Ascochyta* sp. and *Oidiopsis* sp. and to Bean Yellow Mosaic Virus, the quite high proportion of hard seeds and the low percentage in fixed oils and mucilage content are some of its disadvantages.

Typical leaves of these four cultivars are presented in Figure 4.19, while typical seeds of the same cultivars are presented in Figure 4.20. The agronomic and chemical evaluation of these four breeding cultivars are presented in Table 4.4. A detecting pigment paper chromatogram (Figure 4.21) of

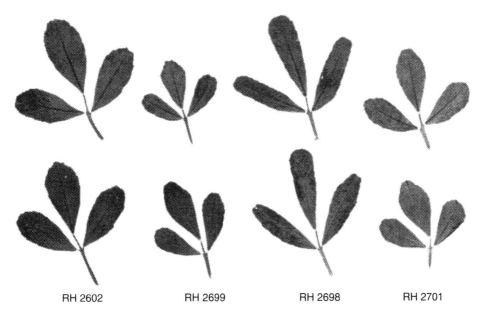

RH 2602   RH 2699   RH 2698   RH 2701

Figure 4.19 Leaves of four breeding cultivars of fenugreek (from left to right: Fluorescent, Ethiopian, Kenyan and Moroccan).

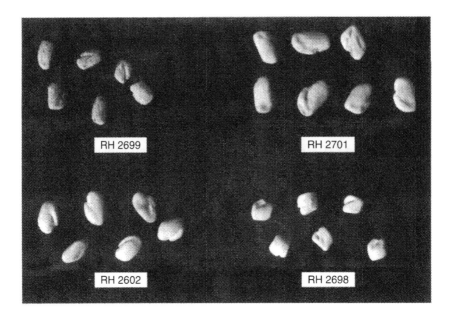

Figure 4.20 Seeds of four breeding cultivars of fenugreek (RH 2602 = Fluorescent, RH 2699 = Ethiopian, RH 2698 = Kenyan, RH 2701 = Moroccan).

Table 4.4 Agronomical and chemical evaluation of four breeding cultivars

Characters	Cultivars			
	Fluorescent	*Ethiopian*	*Kenyan*	*Moroccan*
Agronomic characters				
Seed yield (g/plant)	12.1*	12.0	17.1	15.0
Height (cm)	80	70	75	58
Fertility (ovules fertile %)	95.6	94.8	84.2	97.3
Pods per plant	120.2	121.4	160.4	90.8
Twin pods (% of total)	3.3	8.9	19.8	9.3
Seeds per pod	13.4	13.1	16.4	14.6
Shedding of the seeds (scale $1 \rightarrow 5$)	2.2	3.1	4.1	1.8
Mineral deficiencies (B, Mg, Mn) sensitivity	32.2	26.4	30.4	25.3
Hardness of seed (%)	0	0	40	15
One thousand seed weight (g)	29	26	17	27
Specific seed weight	82–86	78–81	77–79	75–78
G/h index	57	46	32	43
Endosp./Testa ratio	1.8	1.6	1.4	1.3
Chemical evaluation of seed (m.f.b)				
Diosgenin (column/I.R. %)	1.38	1.18	1.51	1.19
Crude protein (%)	30.7	31.8	25.7	30.1
Fixed oil (%)	9.3	9.4	8.4	7.6
Mucilage (%)	21.2	18.9	20.1	17.0

* A high figure indicates that the cultivar shows the character to a high degree.

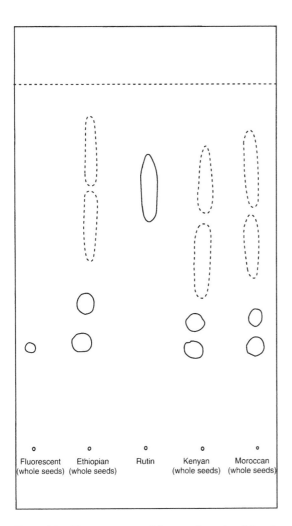

Figure 4.21 Chromatogram of fenugreek seeds of four breeding cultivars, showing the presence of only one colour spot in the Fluorescent cultivar (Solvent: Butanol : Acetic acid : Water 4 : 1 : 5. Visualisation with ammonia).

these cultivars indicates that in the Fluorescent cultivar there is only one pigment and this is in the germ while there is no pigment in its testa (Petropoulos, 1973).

Although the agronomic and other characteristics of a fenugreek cultivar vary greatly between localities, most of them are quite well adapted in the environments in which they are grown. This means the four breeding cultivars are better adapted in northern countries with cold and wet climates.

Comparing the seed yield components among the four cultivars for UK conditions (area of Bath) as seen in Table 4.5 it is concluded that Moroccan and Kenyan were the highest seed yield producers, while Ethiopian and Fluorescent the poorest ones. The superiority of the Moroccan is mainly due to its precocity ensuring a higher percentage of full mature pods while the Kenyan, although it has very small seeds, possesses more pods/plant and seeds/plant than the other cultivars.

Table 4.5 Theoretical seed yield of four breeding cultivars, based on seed yield components by UK conditions (area of Bath)

No.	Seed yield components	Cultivars			
		Fluorescent	Ethiopian	Kenyan	Moroccan
1	Pods per plant	120.2	121.4	160.4	90.8
2	Seeds per pod	13.4	13.1	16.4	14.6
3	Plants/ha	158.480	158.480	158.480	158.480
4	Percentage of pods reached full maturity	31.3	37.2	42.3	53.2
5	One thousand seed weight (g)	29	26	17	27
6	Seed yield (kg/ha)	2.317	2.288	2.998	3.018

Table 4.6 List of some promising genotypes of fenugreek

No.	Genotypes	References	Country of creation	Remarks
1	HFM 1, HFM 7, HFM 8, HFM 13, HFM 14, HFM 17, HFM 18, HFM 19, HFM 20, HFM 22, HFM 25, HFM 27, HFM 29, HFM 30, HFM 34, HFM 37, HFM 39, HFM 54, HFM 61, HFM 63, HFM 65, HFM 116	Paroda and Karwasra, 1975	India	Details in the text
2	RH 3112, RH 3113, RH 3114, RH 3116, RH 3117, RH 3119, RH 3120, RH 3122, RH 3109/32, RH 3109/33, RH 3109/42, RH 3110/66	Petropoulos, 1973	England	Details in the text
3	Green trailing	Singh and Singh, 1974	India	Details in the text

Promising genotypes

The most promising genotypes of fenugreek are summarised in Table 4.6.

The genotypes of Table 4.6 reported by Paroda and Karwasra (1975) were studied for genotype – environment interactions for green fodder yield. Thus, the genotypes HFM 8 and 19 were found to be stable with a high response to changes in environments, while the genotypes HFM 17, 34, 37, 39 and 63 were stable and good for poor environmental conditions. The genotype HFM 39 in particular gave a significantly higher yield over the control and so it was strongly recommended to be included in future breeding programs.

Also, the genotypes in Table 4.6 as reported by Petropoulos (1973) are considered very promising (see Chapter 5) as they were found superior, in comparison with the tested varieties and cultivars, for the following agronomic and chemical composition properties: high yielding (RH 3109/32), resistance to fungus *Ascochyta* sp. (RH 3113, RH 3122), resistance to fungus *Heterosporium* sp. (RH 3114, RH 3120), resistance to winds (RH 3117, RH 3119), high precocity (RH 3114, RH 3116), high diosgenin content of seed (RH 3109/42, RH 3110/66) and high protein content of seed (RH 3109/33). The genotype RH 3112 in particular is an induced mutant and it is very promising and valuable as it is simultaneously high yielding. It is 15 per cent higher in diosgenin content than the mother Kenyan cultivar. It has a short period between the start of ripening and full ripening of pod, has no shedding of seeds and is resistant to winds.

The genotype 'Green trailing', which is a spontaneous mutant, is very promising as it is 30 days earlier in flowering and consequently in maturity than the mother clone 'IC-74' (Singh and Singh, 1974).

Rotation and intercropping

Cropping systems

The cropping or production systems that are usually applied in a fenugreek crop are: (i) The fall and the spring crops and (ii) the pure and the intercropped crops.

As fenugreek is fairly drought resistant (Talelis, 1967; Duke, 1986) and fairly frost resistant (Talelis, 1967; Bunting, 1972) it is generally sown either in the fall and grown as a winter crop in areas with mild winter, or it is sown in spring and grown as a summer crop in areas with soil that keeps moisture during this growing season.

Fenugreek is cultivated either as a pure (unmixed) crop or it is intercropped with other plant species.

Rotation

In the pure crops a rotation system is usually applied where the growing of different crops takes place in a regular order or sequence, and the means of preceding and following crops are used.

The continuous or too frequent growing of different crops results in the rapid breakdown of organic matter and leaves the soil bare and exposed to erosion, while the loss of organic matter also reduces the water absorbing and water holding capacity of the soil.

In choosing a rotation for any given farm or field its relative fertility, the erosion dangers, the diseases, insects or weeds control, the use of equipment, the distribution of labour, the requirements of foods for humans and livestock and finally the achievement of the greatest profit from the farm as a whole over a period of years, must be taken into consideration. As Edison (1995) has reported for spices, including fenugreek, research on crop rotations and the cropping system needs intensification.

Fenugreek is considered a good soil renovator (Duke, 1986). Also, it is very effective in conserving moisture because of its weakly developed root system and it was considered very suitable as green manure in California (Piper, 1947). In contrast all the deep rooted legume crops, such as lucerne or sweet clover, when they are turned under for green manure the soil has usually dried out to a depth of several feet (Arnon, 1972).

The earlier the fenugreek is harvested the higher the amount of residual moisture. It may be said that the most promising approach towards raising the level of soil fertility mainly in the semi-arid regions of sufficient rainfall is the inclusion in the rotation of a fenugreek forage crop and second a fenugreek seed crop. This is because when the fenugreek is cut before the seed is formed the amount of plant nutrients removed from the soil are relatively small, while the soil is enriched in N and organic matter and weeds are cut before flowering and are therefore well controlled. Also fenugreek as a means of increasing soil moisture in dry areas is of overriding importance, and it is very effective for maintaining soil fertility.

As far as the types of rotation are concerned, the areas devoted to a soil improving crop may vary from one-half, one-third, to one-sixth of the total area (Arnon, 1972). Hence, we get a plethora of two-, three-, four-, five- and six-year crop rotations.

So, a very good two year rotation crop is *fenugreek–wheat*, which is widely practised (Dachler and Pelzmann, 1989). The two crops complement each other culturally and nutritionally in the making of different types of bread. The usefulness of fenugreek as a commercial crop is now being recognised and also as a break-crop for cereal areas (Hardman, 1969). The introduction of fenugreek into rotation rapidly restores the productivity of worn-out wheat soils.

As fenugreek requires a field free of weeds (Anonymous, 1987) it must not be sowed after a crop that is favourable to the growth of weeds or that may destroy the texture of the soil (Anonymous, 1987), and because it is a legume that fixes the atmospheric N, it could follow a high N consumer crop like tobacco (Anonymous, 1987; Heeger, 1989). So, a successful three-year crop rotation system could be *cereal–fenugreek–tobacco*.

A specific rotation usually has one or more cultivated or row crops, at least one small grain crop and a sod crop (Hughes and Henson, 1957). Thus, an effective four-year rotation could be *potatoes–wheat–fenugreek–tobacco*.

It appears that the crop yield can be economically maintained regardless of the rotation, providing it contains a legume like fenugreek at least once in 5 years, and the crops are reasonably fertilised (Hughues and Henson, 1957). So, a very good five-year rotation could be *corn–potatoes–wheat–fenugreek–tobacco*.

All these rotation systems are based on observations, adaptation trials and discussions among the scientists, who conducted the trials in many countries. The influence of these crop rotations on crop yields and other effects of certain crops on succeeding ones has been reported from most of the corresponding agricultural experiment stations. For example, it has been estimated that in rotation the effect of fenugreek as a preceding crop on the amount of residual moisture in soil (30–120 cm depth) at time of sowing, and of the following wheat crop, was: fenugreek for green-manure 3,500 m^3/ha and fenugreek seed 2,520 m^3/ha (Arnon, 1972).

It should be emphasised that there are many different rotations including fenugreek in use and there may be several different rotations that will give equally satisfactory results for any one soil and climatic condition.

Intercropping

When fenugreek is intercropped it is used either as a main crop or as first, second etc. intercrop.

Fenugreek, as an erect crop, offers support in the inter-creeping legumes (Talelis, 1967), while in the case of fenugreek for forages the intercropping reduces its peculiar smell that causes tainting in milk and meat and their derivatives (Talelis, 1967; Duke, 1986). Fenugreek for forages is intercropped in Greece with vetch, faba beans, horse bean (Talelis, 1967; Dalianis, 1987), barley and clover (Talelis, 1967) and in Europe with alfalfa (Talelis, 1967) and faba beans (Heeger, 1989). Fenugreek for seed is intercropped in India with coriander, gingelly, bengal gram (Duke, 1986), turmeric (Sekar and Muthuswami, 1985) and sugarcane (Singh and Rai, 1996), and in Poland with anise (Anonymous, 1987). It must be emphasised that when turmeric (main crop) was intercropped with fenugreek (first intercrop) the highest net income was obtained in comparison with other applied combinations of turmeric intercropping (Sekar and Muthuswami, 1985).

Harvesting

Maturation

The maturation of the fenugreek plant, especially of the pod, should be studied before the examination of the plant harvesting itself. It is known that fenugreek has an indeterminate, growth habit continuing from the terminal and axillary buds, while flowering and formation are even and maturation of pods are still in progress. The process of fenugreek maturation depends to some degree upon conditions external to the plant, such as, first climate and season, second the natural character of the soil and third on artificial factors concerned with cultivated practice. Fenugreek seed development lasts approximately 120 days after anthesis (Campbell and Reid, 1982).

Stages of pod development

It was found that the pod of fenugreek has the following four distinguishing stages of development (Petropoulos, 1973):

1   *First stage (Length development)*. It starts from the time of flower fertilisation until the pod takes its approximately final length, but it is still narrow. The seeds are very small. There is no differentiation between testa and endosperm of seed, while the germ is invisible. Duration of this stage: 25–30 days.

2   *Second stage (Width development)*. The pod is chiefly developed in width. The husk of the seed increases characteristically and the differentiation between testa and endosperm is clear, but the weight quotient endosperm/testa is smaller than 1. The germ is visible now but the ratio of germ/husk is also smaller than 1. Duration of this stage: 20–25 days.

3   *Third stage (Germ development)*. The size of the pod remains nearly the same, while the germ (embryo) of the seed increases characteristically and the quotient of germ/husk starts to become higher than unity. The husk increases slowly but the ratio of endosperm/testa starts to become higher than unity. This stage takes 35–45 days.

4   *Fourth stage (Ripening)*. The pod, according to the variety, changes colour from green to light straw for some varieties and from green to light brown for others, starting from the tip of the pod to the base. At the same time, the embryo of the seed changes colour from green to yellow for some varieties and from green to purple for others, starting from the area of the radicle. Pods open slowly, hence the fenugreek crop is easily handled for seed production (Duke, 1986). But in some cases the shattering of pods and scattering of seeds takes place especially in some varieties, which are sensitive from this aspect. Duration of this stage is from 15 to 20 days.

It was found that when the first lower pod of each shoot is completely ripe, the pods up to the fifth or sixth node have started to ripen and their maturation, especially for experimental purposes, could be continued artificially at room temperature for about 3 weeks (Petropoulos, 1973).

Harvesting for forage

Although the use of fenugreek for forage is very limited today, mainly because of its peculiar smell that causes tainting of animal products and their derivatives (Molfino, 1947; Talelis, 1967; Dalianis, 1987), it is still used in India and Turkey as green fodder and hay for cattle. Hardman (1997) suggests it can be used as an alternative to lucerne or forage peas, while he confirms its use as silage in Japan, and says that fenugreek seed and straw are shown to be superior to other legume seeds and straws in a balanced feed with sheep *in vivo* experiments in Spain.

Time of harvesting

It is evidently important to harvest fenugreek crop for forage at the time that will allow the greatest yield, and at the same time ensure a product of high quality.

The losses resulting from the delayed harvesting of forage fenugreek are due to the shattering of the leaves, reduction of palatability and decrease in nutritive value, while the disadvantages of its premature harvesting are the lower yields of food constituents and dry matter and the greater difficulty in curing.

Hence, according to all indications, the best harvesting time of fenugreek for green fodder should be when the pods of the base are in the first stage of their development, where the plants are a well formed mass and are very tender. For hay these pods complete the second stage of their development and the germ of the seed that is rich in protein has started to increase in size, which at the same time increases the protein content of the hay.

Harvesting in a manner to save the most leaves with the stems is a primary goal. In India, for forage, October plantings are cut in February–March, while when sown in January the plantings are cut in April (Duke, 1986).

There is a theory that the responsibility for the peculiar odour of fenugreek plant is the alkaloid trigonelline (Marques de Almeida, 1940), during germination and growth the trigonelline content varies and a marked decrease is noted during the first 30 days followed by a regular increase up to seed formation. This may be taken into consideration as far as the time of harvesting is concerned, although Molfino (1947) believes that this odour is due to the contented coumarin.

Methods of harvesting

Fenugreek can be cut and handled either by labourers or with ordinary farm cutting equipment or by conventional mowers, conditioners and rakes. Also, the use of rectangular balers and forage harvesters are recommended for special farm situations in the future.

Drying of hay

The moisture reduction of the fenugreek hay from up to 75 per cent to less than 12 per cent for storing, constitutes one of the most difficult of all crop harvesting jobs. In a warm dry climate, drying is affected by simply exposing the cutting hay to the air in shallow layers. But under humid conditions, in most cases, the fresh hay has often been oven-dried.

Mode of use

When fenugreek is used as forage it is mostly harvested as hay. Under adverse climatic conditions, however, it is often saved as brown hay. It may also be preserved as silage but this is seldom done, except when weather conditions prevent drying and a silo is available. Fenugreek is also used as fodder as a sort of straw, that is, after the seed has been threshed, but its palatability is quite low.

Harvesting for seed

Fenugreek is cultivated mainly for seed production (Piper, 1947; Hidvegi *et al.*, 1984; Dalianis, 1987). So, the harvesting for seed presents special interesting information.

Time of harvesting

Mature pods on the lower part of the plants are usually ready to be harvested, while new flowers and pods are still forming at the top because of its indeterminate growth habit. The decision as to when to harvest, is always arbitrary. Harvesting too late permits ripe pods to shatter and scatter and the seeds are lost, while harvesting too early means an excessive amount of unripe pods with green shrivellent seeds. So, in most of the cases, especially under wet conditions, harvesting starts when most of the pods are mature (Anonymous, 1987; Heeger, 1989).

Fenugreek ripens usually *c*. 3–5 months after planting (Fazli and Hardman, 1968), but this is true for spring sowing, as in Greece for fall sowing in November this time exceeded 7 months. In Poland for sowing in April with fine weather, the harvest is in August – beginning of September (Anonymous, 1987).

Methods of harvesting

Two methods of harvesting fenugreek seed are usually employed: the *traditional* method, where the plants are cut and handled mainly by labourers, and the modern one of *mechanical* harvesting.

According to the first method, the plants are cut by labourers with ordinary farm equipment when most of the pods are mature. The uprooted plants are left in the field to dry for a few days until the maturation of the green pod seeds especially those of the nearly mature. Then they are thrashed with a grain thrasher, winnowed (by wind sifting), further dried and stored. This method, which is still applied by some farmers in underdeveloped fenugreek producing countries, is laborious and is characterised by a high cost of production and a high percentage loss of seed. This is because of the shattering of pods and scattering of seed, mainly from the many removals of the plants.

Mechanical harvesting usually has two versions: the *direct combining* and the *windrow curing* and then threshing with a pick up combine.

Direct combining is applied in countries with a dry climate and in fields with low moisture content soils and winds apt to disturb windrows. An advantage is that the harvest can be delayed until nearly all the pods are ripe, but not too delayed, as extensive losses can result from shattering. Direct combining should start as soon as the pods and the leaves are dry (15–20 per cent moisture) even though the stems are still relatively green (40 per cent). The seed should be aired within 24 h of harvest.

It is necessary to adjust the combining machine for seed harvesting to prevent heating, as the first pods of the base are very close to the surface of the soil and also to adjust the auger speed, usually reducing it below what has been recommended by the manufactures.

Windrow curing is preferable where fields are late maturing with a high proportion of green pods caused by high soil moisture, as the maturation can be continued in the windrow especially with the nearly mature seeds. A conventional mower should cut plants when up to 80 per cent of the seed pods are mature. Windrow curing follows for a few days until the maturation of most of the green pod seeds and then threshing with an adapted pick-up combine used for harvesting seed legumes (preferably for lentil), when the moisture content of the foliage is from 12–18 per cent. In Poland (Anonymous, 1987) the uprooted plants are put in a truss obliquely like a cone on the soil, or on a three-legged wooden skeleton to dry and then they are threshed. Windrow losses could be higher if winds in the area are troublesome.

In both of these versions mechanical injury to the seed from improper combining adjustment may cause losses and a deterioration in the yield. It is important to check each combine for mechanical damage of seed when harvesting begins. If the percentage of visible injury exceeds 5 per cent, the combine should be stopped and necessary adjustments should be made to minimise the seed damage.

Harvesting in wet climates

In wet climates and generally in prolonged wet weather during the harvesting period of fenugreek, many problems are created (Petropoulos, 1973; Jorgensen, 1988). In these cases, the peculiarity of indeterminate growth of fenugreek becomes intensely obvious with the continuous regrowth of the plant, resulting in the simultaneous presence of ripe pods on the base and complete unripe ones on the top. So, seeds germinate while they are still in the pods on the plant, especially in sensitive cultivars like Fluorescent, and shattering on the pods on the base and scattering of the seeds occur because of the prolonged growth of the plants. Under these wet conditions, in order to help the full maturation of the seed and to avoid the above losses the following actions may be necessary before threshing:

i   to stop any late feeding by N;
ii   to sacrifice the pods of the upper top of the stem by top cuttings in order to obtain timely harvesting of the rest and reduce shattering losses; and
iii to apply a desiccant chemical, like reglone, dinoseb, diquat, etc., which should be used according to the instructions from the manufacturer and based on local experience, and left for at least ten days before direct harvesting.

Despite all these actions, it was found (Petropoulos, 1973) that in England by natural ripening, the proportion of the seeds per plant that reach full maturity varies according to the cultivar and the average is only 40 per cent. But we must not ignore the fact that under these wet conditions the potential productivity of fenugreek is very high in comparison with the non irrigated crops of traditional fenugreek producing countries. But in very heavy wet conditions it is often difficult to avoid the oven-drying of the raw material to 10 per cent moisture content, before threshing.

Drying

As had been reported previously, in warm dry climates drying is effected by simply exposing the cut plant material to the air in shallow windrows, or in an oblique truss on the soil, or in rainy weather on a three-legged wooden skeleton. But under very heavy humid conditions, the plant material has nearly always been oven-dried, especially for valuable experimental stock.

The temperature and the total time of drying are important. Generally for seed production the temperature of drying air should not exceed 32–43°C (Anonymous, 1961). It was found for the production of certified seed of fenugreek, especially for some sensitive cultivars (i.e. Kenyan and Moroccan), that the temperature should not exceed 35°C to limit the proportion of hard seeds (see Table 4.2). Also, drying in high temperatures, sets up stresses between the inner and outer areas of the fenugreek seed particularly at high moisture levels, which can result in the severe cracking of the seed coat, especially in the sensitive Fluorescent cultivar (Petropoulos, 1973).

It is important to find the proper balance between too rapid drying with resulting coat cracking or coat hardening and preventing complete drying and too slow drying with deterioration of the seed, especially under bad ventilation conditions. So, the total drying time for any seed is influenced by its initial and final moisture content, its drying rate, the rate of airflow and the temperature of the drying air.

The final moisture content for safe storage of seed is generally 4–14 per cent, depending on (i) the kind of seed, that is, for fenugreek it has been estimated to be 10 per cent (Petropoulos,

1973), while the appropriate Polish Institute (Anonymous, 1987) reports 11 per cent, (ii) the type of storage and (iii) the anticipated storage period. It must be said that fenugreek seeds retain their viability for many years (Petropoulos, 1973; Duke, 1986). Lower moisture levels are generally desirable for longer storage time and confined storage conditions.

A rule of thumb that can be used to determine drying time is that about 0.3 per cent of the moisture can be removed per hour with an air flow rate of 11.5 m^{-3} per minute per ton (m^3/min · ton) at 43°C (Anonymous, 1961). This drying rate varies with different seed, temperature and initial moisture. The hourly rate will be less if the initial moisture content is low and if the drying air is unheated or is at temperatures below 43°C.

It was found, by England conditions, that the fresh harvested raw material of fenugreek (stem with pods) at the stage of approximately 20 per cent ripe pods, contains 65–70 per cent of moisture that is distributed among the different plant parts in the following proportion: stems and leaves 47 per cent of total moisture (weighting 42 per cent of the total weight with a moisture content of 73 per cent) and pods including seeds 53 per cent of the total moisture (weighting 58 per cent of the total weight with a moisture content of 59 per cent) (Petropoulos, 1973).

Also, it was found that every time 11.4 bushels of this raw material was dried with good results to a final moisture content of around 10 per cent in an electric oven volume 130 c.f. with air intake 20°C, oven air temperature around 36°C and air flow rate 57.2 c.f.m., in a drying time of 182 h (Petropoulos, 1973).

Cleaning

Fenugreek after threshing and collecting should be cleansed of the extraneous matter and the other impurities by a suitable seed cleaning machine. There are many types of seed cleaning machines that operate on the basis of size, shape, density and surface texture. In Greece adapted wheat seed cleaning machines are used for cleaning fenugreek seeds.

A suitable cleaning system should permit efficient handling of seeds, prevent injury to them, avoid mixtures and maximise return from labour and supervisory personnel. The method of handling, whether in bulk, sacks or both will influence the overall design.

After the seed cleaning, the threshold of quality standards that should not be exceeded for a first and second class quality seeds of fenugreek, according to the appropriate Polish Institute (Anonymous, 1987), as far as the purity is concerned are respectively: (i) extraneous organic matter: 2 and 3 per cent, (ii) extraneous mineral matter: 0.5 per cent in both cases, (iii) other parts of the plant: 1 per cent and 5 per cent, (iv) seeds with different colour: 5 per cent and 10 per cent and (v) matter that goes through from a sieve 1.6 mm: 3.5 per cent, in both cases.

Storage

Special care is needed mainly for the storage of the fenugreek seeds, as the storage of hay is an easy story. The distinguishing of storage of seeds for common use and of seeds for seeding is necessary. In the first case interest presents the preservation of seed, while in the second case the primary purpose is to retain their viability and vigour for many years.

Several factors may determine the healthy situation and longevity of fenugreek seeds stored in a natural environment, like moisture, temperature, seed coat character, maturity and insect infestation. For best results fenugreek seeds must be stored in an environment with less than 10 per cent moisture with a temperature near 0°C.

Fenugreek seeds retain their viability for long periods. In Greece, fenugreek seeds forgotten in a truck for 47 years germinated very well. There are some indications for the Fluorescent cultivar that the viability of its seeds is reduced rapidly in comparison with other cultivars, this may be due to the homologous pair of recessive genes that control the lack of colour and the natural splitting or crazing (cracks) of the seed coat, which often appears in this cultivar (Petropoulos, 1973). It has been reported that certain homologous recessive characters are related to the reduction of vitality of corn seeds (Anonymous, 1961). Also the Fluorerscent cultivar possesses soft instead of hard seeds and according to Mercer (1948) the hardness of the seed coat protects the viability of seeds.

Yield

Fenugreek as a cultivated crop, as has been reported previously, is grown and harvested principally for the seeds and secondarily as forage.

Seed yield

Fenugreek, as a legume crop, produces its seeds in pods. So, seed productivity is related to the yield components that include at the time of harvest (i) seeds/pod, as an average of the variety, (ii) pods/plant, as an average of the variety, (iii) the proportion of pods that reach the full ripe stage, mainly according to climate and weather conditions at harvesting time, (iv) one thousand seed weight and (v) plants/unit area according to the applied plant density.

These yield components for the average of four breeding cultivars by English conditions (Petropoulos, 1973) and the Ionia variety in Greek conditions (personal experience) are presented in Table 4.7.

As different varieties of fenugreek are cultivated in different conditions throughout the world, a wide range of seed yields have been reported by various authors. So, Banyai (1973) reported that in India from twenty-nine ecotypes of fenugreek tested, seed yields are 500–3,320 kg/ha and that yields of 1,800 kg/ha were economically viable, while the average seed yield of the last twenty years (1975–95) in India is 1,203 kg/ha (Anonymous, 1996a). Mohamed (1990) reported a seed yield of 1,595 kg/ha in Egypt, Piper (1947) reported 1,680 kg/ha in USA, while Talelis (1967) estimated the seed yield in Greece as 2,465 kg/ha. In Ethiopia, the seed yield for

Table 4.7 Yield components for different varieties and various environmental conditions

No.	Yield components	Variety: the average of the four breeding cultivars (Fluorescent, Ethiopian, Kenyan, Moroccan) environment: wet and cold	Variety: Ionia environment: dry and hot
1	Seeds/pod (Average)	14	10
2	Pods/plant (Average)	123	48
3	Plants/unit area (N/ha)	158.480	158.480
4	One thousand seed weight (kg)	0.025	0.018
5	Percentage of full ripe pods	40.5	95
6	Seed yield/unit area (kg/ha)	2.763	1.300

fenugreek was presented as very low, fluctuating between 582 and 608 kg/ha (Anonymous, 1970), while in Poland (Anonymous, 1987) this value was 495–1,480 kg/ha and in Germany 1,700–2,100 kg/ha (Dachler and Pelzmann, 1989). In England, a seed yield of 3,700 kg/ha has been reported from experimental fields (Petropoulos, 1973; Evans, 1989).

Forage yield

Fenugreek has long been recognised as good forage, especially in ancient times where the species takes the name *foenum-graecum* that means 'Greek hay'. But in modern times this use has dwindled greatly and other forages have replaced it (Pantanelli, 1950; Rouk and Mangesha, 1963).

As different varieties of fenugreek are cultivated in different conditions and are cut for forage at different stages of growth, a broad range of forage yields has been reported. So, Piper (1947) reports that the yield of fenugreek as fresh matter was estimated to be 13,170 kg/ha at Santa Paolo of California and 17,400 kg/ha in San Joaquin Valley, while Duke (1986) reported that according to the Wealth of India the green forage production of fenugreek is estimated at 9–10 M.T./ha. Paroda and Karwasra (1975) studying twenty-four genotypes, reported that forage dry matter yields about 1,500–2,750 kg/ha with a mean of *c.*20,000 kg/ha, while Heeger (1989) reports a green hay yield of 2,000 kg/ha and for dry hay 5,000 kg/ha. The straw production of fenugreek in Greece is estimated at 1,850 kg/ha.

Uses

Fenugreek is a chemurgic cash crop, usually cultivated as a break crop for cereal, as it is considered a good soil renovator. The whole plant is used as forage and vegetable, while the seeds (whole, powdered, in flour, or roasted) are used as human and animal food, spice, dyeing, flavouring, as well as for medicinal and industrial purposes.

Animal food

Originally, it was grown in the ancient world and especially in Europe and was recognised as a good forage, hence the name 'Greek hay' or *foenum-graecum* (Rouk and Mangesha, 1963). In India and Turkey it is used as green fodder and hay for cattle. Hardman (1997) suggests it as an alternative to lucerne or forage peas, while in Japan, according to this researcher, it is used as silage. Mildewed or 'sour' hay is made palatable to cattle when fenugreek herbage is mixed with it. Also Hidvegi *et al.* (1984) report that fenugreek seeds are used for feeding cattle. Ground fine and mixed with cotton seed it is fed to cows to increase the flow of milk. An extract of fenugreek seed is added to animal food to increase its palatability (Smith, 1982), for example, when powdered mineral magnesite is added to cattle feed to maintain milk production or when the feed requires it (see section on 'Flavour extracts').

But in modern times other forages have replaced fenugreek (Pantanelli, 1950; Rouk and Mangesha, 1963). In the Middle Ages it is recorded that it was added to inferior hay because of its pleasant but peculiar smell (Howard, 1987). Molfino (1947) and Talelis (1967) notice that fenugreek hay causes the tainting of milk and its derivatives. Also Duke (1986) reports that fenugreek increases the flow of milk in cows but impacts its aroma. According to our observations and experience, if the flavour is unwanted in the meat then fenugreek fodder should be discontinued several weeks before slaughter (Petropoulos, 1973), while Hardman (1997) suggests that in order to avoid tainting of milk and meat it should be withdrawn from the diet 3 weeks before milking or slaughter.

Human food

Young plants and fresh tips of fenugreek are succulent and eaten as a salad, or cooked and generally served as a condiment in India and Egypt, as the fresh plant is very rich in vitamin C
(207 mg per cent) (Saleh *et al.*, 1977).

The fenugreek seed is rich in protein, fixed oils and minerals and so it is nutritive and a tonic
(Anonymous, 1994). It is an important fodder crop for those countries in the Middle and Far
East where meatless diets are customary for cultural and religious reasons. Fenugreek protein is
rich in lysine (345 mg/g) and in comparison to the data for human requirements, calculated
from the amino acid pattern, approaches that of soybeans (Hidvegi *et al.*, 1984). Fenugreek contains *c.*5 per cent oil with a strong celery odour and is used in butterscotch, cheese, licorise,
pickle, rum, syrup and vanilla flavours (Duke, 1986). It is supposed to stimulate the appetite
(Parry, 1943) and the digestive process (Fazli and Hardman, 1968). Egyptians and Hindus cultivated it for food (Howard, 1987). In Sudan and Egypt the seeds are used in making beverages
and in some countries the roasted seeds are used as a coffee substitute, probably because of the
alkaloid trigonelline content, which is a basic constituent of the coffee seed. While in Ethiopia
the seeds are prepared for infant feeding by boiling the whole seed (Fazli and Hardman, 1968),
in North Africa it is mixed with breadstuff (Manniche, 1989); in Egypt also the seeds of the
fenugreek are added to bread as a supplement of wheat and maize (Hidvegi *et al.*, 1984). In
Yemen it is widely used everyday by the general population. Fenugreek was considered a warming herb and poor people used it to gain weight, probably because of the high fat content of the
seed (Manniche, 1989). Harem women were said to consume roasted fenugreek seeds to attain
buxomness (Duke, 1986), while William and Thomson (1978) report that the seeds cause an
alluring enhancement and roundness of the breast. Sprouting seeds are used as vegetables
(Stuart, 1986).

Spice

As a spice, fenugreek seeds add nutritive value to food, as well as flavouring and are used in
soups and curries (Duke, 1986). In the UK and the US it is used in the manufacture of chutneys
and various spice blends, for example, in some curry powders (Rosengarten, 1969). Fenugreek
seed is commonly used for seasoning purposes and as an ingredient of curry powder and sauces
(Fazli and Hardman, 1968). In Greece and Turkey with seed powder and beef, it is used to make
the bacon 'pastrumas' (Petropoulos, 1973; Dalianis, 1987), while in the Middle East with fenugreek seed powder and other ingredients the confectionery 'halva' is made (Stuart, 1986).

Repellent–flavouring–perfume

Fenugreek, as most of the species of the genus *Trigonella*, is strongly scented (Anonymous, 1994)
and serves as an insect repellent (Duke, 1986). Chopra *et al.* (1965) report that in the Punjab
district of Pakistan they mix the dry plant of fenugreek with grains in order to protect them
from attacks of insects, particularly during the rainy season. In Turkey fenugreek seed is
placed between cloths to repel cloth moths, while Evans (1989) reports that leaf extracts repels
numerous common insects.

The main use of the imported quantities of fenugreek seeds from countries of Europe and
America is the extraction of a flavour liquid (Smith, 1982). This flavour extract in the USA and
Canada is used mainly as an artificial imitation of maple syrups, in tobacco flavours and some
spice seasonings, while in Europe (UK, Germany, Netherlands, Belgium, etc.) its main use is in

animal feed flavours and secondarily in food flavours (Bread, cheese, tea, pizza, etc.) (Smith, 1982).

The seeds of the fenugreek are well known for their pungent aromatic properties (Max, 1992). The aroma of the fenugreek volatile oil is strong, sweetish, pleasantly bitter and reminiscent of burnt sugar (Anonymous, 1982) and it also possesses a strong smell of goats (Schauenberg and Paris, 1990), while its main constituent is the 3-hydroxy-4.5-dimethyl-2(5H)-furanone (Girardon *et al.*, 1986). Its aroma may in fact be the secret of a very successful french perfume (Igolen, 1936; Fazli and Hardman, 1968).

There is a contradiction regarding the origin of the peculiar smell of fenugreek that causes the tainting of animal products. So, according to Marques de Armeida (1940) it is due to alkaloid trigonelline, while Molfino (1947) supports that the contained coumarin is responsible for this peculiar smell.

Dyeing

The fenugreek seed contains a yellow dye that is used for dyeing cloth and could be used for other colouring purposes, including possibly food and pharmaceutical products. This dye, when mixed with copper sulphate, produces a fine permanent green (Fazli and Hardman, 1968). The same workers report the use of fenugreek in the preparation of imitation carmine.

Remedy

Fenugreek seeds have been known and valued as medicinal material from very early times. Fenugreek was widely cultivated as a drug plant (*semen foenugraeci*) until the nineteenth century. The mucilaginous seeds are reputed to have many medicinal virtues, as a tonic, emollient, carminative, demulcent, diuretic, astringent emmenagogue, expectorant, restorative, aphrodisiac and vermifugal properties and were used to cure mouth ulcers, chapped lips and stomach irritation (Duke, 1986). When soaked in water, the seeds swell and produce a soothing mucilage said to aid digestion (Fazli and Hardman, 1968; Rosengarten, 1969). The decoction is given to strengthen those suffering from tuberculosis or recovering from an illness (Lust, 1986). Also the decoction is used for gargling for sore throat and internal inflammation of the stomach, intestines and ganglia (Schauenberg and Paris, 1990). Crushed seeds with powdered charcoal are used to make a hot mushy for external use in cataplasms, ointments and plasters, applied to bruises, swellings, boils and ulcers (Potterton, 1983; Bunney, 1984), like the swelling of testicles (Reger, 1993). As the seeds contain up to 50 per cent of mucilaginous fibre they have been used internally because of their ability to swell and relieve constipation and diarrhoea (Evans, 1989; Sharma *et al.*, 1996). A poultice of seeds is used for gouty pains (Sharma *et al.*, 1996), neuralgia, sciatica, swollen glands, wounds, furncless, fistulas, tumours, sores, skin irritation, abscesses and carbuncles (Potterton, 1983). Fazli and Hardman (1968) report that a decoction of the seed is taken in East Africa as a remedy for gonorrhoea, a former use in European medicine and a poultice of seeds as a local remedy for vermin. In Malaya, they poultice the seeds onto burns and use them for chronic coughs, dropsy, hepatomegaly and splenomegaly (Duke, 1986; Bhatti *et al.*, 1996; Sharma *et al.*, 1996). The Chinese use the seed for abdominal pain, chilblains, cholechystosis, fever, hernia, impotence, hypogastrosis, nephrosis and rheumatism (Duke, 1986). Fenugreek tea is mucilaginous, nutritious, and soothing to the intestinal canal (Potterton, 1983). Fenugreek also has been reported as a lactogogue and a spermicidal (Duke, 1986). Externally cooked seeds with water into a porridge, can be used as hot compresses on boils and abscesses in a similar manner to the usage of linseed (Fluck, 1988). As a coarsely ground powder the seeds make a soothing, quietening and convalescent drink (Ceres, 1984). Aqueous and alcoholic extracts have been reported to have a stimulating effect on the isolated guinea pig uterus, especially during the last period of pregnancy, indicating that those extracts may have a high oxytocic activity (Leung, 1980). It has been renowned for expelling poisons and unwanted materials from the human body (Howard, 1987). In India the seeds are used to form the base of a medicinal confection called 'Luddoo' (Rouk and Mangesha, 1963). One report in Java indicates that the seeds were used to prevent baldness but it is not clear as to the nature of the treatment, whether one should eat the seeds or wear them as poultice (Leung, 1980). Externally, the seeds are an emollient and accelerate the healing of suppurations and inflammations (Fluck, 1988).

Aqueous extracts of seeds in Pakistan showed antibacterial activity against a series of bacteria (Bhatti *et al.*, 1996). In veterinary medicine the seeds are used to increase milk production (Bunney, 1984). In Greece and elsewhere in recent times the decoction of the seed is taken as a remedy for diabetes (Evans, 1989; Khosla *et al.*, 1995; Sharma *et al.*, 1996), while in Israel it is used as an oral insulin substitute (Oliver-Bever, 1986). As the fenugreek seed contains very little starch and the polysaccharides are present in the form of silicon-phosphoric ester of

Table 4.8 Recapitulation of the reported therapeutical properties of fenugreek

No.	Therapeutical and pharmacological properties and activities	References
1	Antibacteric	Bhatti *et al.*, 1996
2	Antidiabetic	Evans, 1989; Khosla *et al.*, 1995; Sharma *et al.*, 1996
3	Antihelminthic	Fazli and Hardman, 1968
4	Antidiarrhoeal	Fazli and Hardman, 1968
5	Antihepercholestrolaimic	Vallette *et al.*, 1984; Oliver-Bever, 1986; Sharma *et al.*, 1991
6	Antipyretic	Duke, 1986
7	Antitumour	Singhal *et al.*, 1982; Evans, 1989
8	Aphrodisiac	Fazli and Hardman, 1968; Duke, 1986
9	Astringent	Duke, 1986
10	Carminative	Duke, 1986
11	Convalescent	Ceres, 1984
12	Coughing (ease)	Duke, 1986; Bhatti *et al.*, 1986; Sharma *et al.*, 1996
13	Demulcent	Duke, 1986
14	Digestive	Fazli and Hardman, 1968; Rosengarten, 1969
15	Diuretic	Duke, 1986
16	Emmenagoque	Duke, 1986
17	Emollient	Duke, 1986; Fluck, 1988
18	Expectorant	Duke, 1986; Howard, 1987
19	Galactagoque	Bunney, 1984; Duke, 1986
20	Hypocholesterolaemic	Vallette *et al.*, 1984; Sharma and Ragharam, 1991
21	Hypoglycaemic	Khosla *et al.*, 1985
22	Insulin substitute	Oliver-Bever, 1986
23	Ionic neutral	Duke, 1986
24	Oxytocic	Leung, 1980
25	Restorative	Duke, 1986
26	Spermicidal	Duke, 1986
27	Stomachic	Duke, 1986; Schauenberg and Paris, 1990
28	Suppurative	Fluck, 1988
29	Tonic	Fazli and Hardman, 1968; Duke, 1986
30	Vermifugal	Duke, 1986

Table 4.9 Human and animal diseases or disorders that have been reported as cured by using fenugreek, as a remedy

No.	Reported as cured diseases or disorders	References
1	Abdominal pain	Duke, 1986
2	Absesses	Potterton, 1983; Fluck, 1988
3	Baldness	Leung, 1980
4	Boils	Potterton, 1983; Bunney, 1984; Fluck, 1988
5	Bruises	Potterton, 1983; Bunney, 1984
6	Carbuncles	Potterton, 1983
7	Chilblains	Duke, 1986
8	Chapped lips	Duke, 1986
9	Cholecystosis	Duke, 1986
10	Chronic cough	Duke, 1986; Bhatti *et al.*, 1996; Sharma *et al.*, 1996
11	Constipation	Evans, 1989; Sharma *et al.*, 1991
12	Convalescence	Ceres, 1984
13	Diabetes	Evans, 1989
14	Diarrhoea	Evans, 1989; Sharma *et al.*, 1996
15	Dropsy	Duke, 1986; Bhatti *et al.*, 1996; Sharma *et al.*, 1996
16	Dyspepsia	Duke, 1986; Sharma *et al.*, 1996
17	Fibromas	Singhal *et al.*, 1982; Evans, 1989
18	Fever	Duke, 1986
19	Fistulas	Potterton, 1983
20	Furunculosis	Potterton, 1983
21	Gaglia	Schauenberg and Paris, 1990
22	Glands	Potterton, 1983
23	Gonorrhoea	Fazli and Hardman, 1968
24	Gouty pains	Sharma *et al.*, 1996
25	Hepatomegaly	Duke, 1986; Bhatti *et al.*, 1996; Sharma *et al.*, 1996
26	Hernia	Duke, 1986
27	Hypercholesterolaemia	Vallette *et al.*, 1984; Oliver-Bever, 1986
28	Hypogastrosis	Duke, 1986
29	Impotence	Duke, 1986
30	Inflamations	Fluck, 1988
31	Intestines	Potterton, 1983; Schauenberg and Paris, 1990
32	Mouth ulcers	Duke, 1986
33	Nephrosis	Duke, 1986
34	Neuralgia	Potterton, 1983
35	Recovering from an illness	Lust, 1986
36	Rheumatism	Duke, 1986
37	Scatica	Potterton, 1983
38	Skin irritation	Potterton, 1983
39	Sores	Potterton, 1983
40	Splenomegaly	Duke, 1986; Bhatti *et al.*, 1996; Sharma *et al.*, 1996
41	Stomach irritation	Duke, 1986; Schauenberg and Paris, 1990
42	Suppurations	Fluck, 1988
43	Swellings	Potterton, 1983; Bunney, 1984
44	Throat sore	Schauenberg and Paris, 1990
45	Tuberculosis	Lust, 1986
46	Tumours	Potterton, 1983
47	Ulcers	Potterton, 1983; Bunney, 1984
48	Uterus	Leung, 1980
49	Vermin	Fazli and Hardman, 1968
50	Wounds	Potterton, 1983

manogalactan, which is not hydrolysed by ptyalin or pancreatic amylase (Kamel, 1932) and it may be related with anti-diabetic activity. Fenugreek seeds possess hypocholesterolaemic effects as it reduces serum cholesterol in animals under laboratory conditions (Valette *et al.*, 1984; Oliver-Bever, 1986; Evans, 1989; Sharma *et al.*, 1991). A French patent has been granted to a product purported to have anti-tumour activity, especially against 'fibromas' (Singhal *et al.*, 1982; Evans, 1989). Also, crushed leaves are taken internally for dyspepsia (Duke, 1986; Sharma *et al.*, 1996).

The recapitulation of the therapeutical properties of fenugreek are presented in Table 4.8, while the human and mainly animal diseases that have been cured by using, are listed in Table 4.9.

Industrial material

Fenugreek as a chemurgic crop has a wide use for industrial purposes. Its seeds are considered to be of commercial interest as a source of a steroid diosgenin, which is of importance to the pharmaceutical industry as a starting material in the partial synthesis of corticosteroids, sex hormones and oral contraceptives (Fazli and Hardman, 1968; Hardman, 1969; Khanna *et al.*, 1975; Kiselev *et al.*, 1980).

After diosgenin extraction a series of side-products like protein, fixed oils, oleoresin (coumarin, mucilage, gums) might be extracted (Duke, 1986). These by-product residues may be used for organic (biomass, fuels, manure) and inorganic (chemical fertilisers as fenugreek seeds are rich in N and potassium) purposes.

The husk of the seed may be removed for its mucilage with the remainder partitioned into oil, sapogenin and protein rich fractions (Duke, 1986), while the oil can be used in food and soap industries and also as a galactogogue (Fazli and Hardman, 1968).

Seed polysaccharide mucilage (galactomannan), about 25 per cent, could be prepared from the mark left after the extraction of fixed oils (used as a lactagogue). Its relatively high viscosity makes it a good emulsifying agent to be used in the pharmaceutical and food industries. Due to its neutral ionic properties it is comparable with other drugs or compounds sensitive to acids (Duke, 1986). Efforts have been made to identify the mechanisms of fenugreek galactomannan biosynthesis (during seed development) and hydrolysis (during germination) in order to produce transformed fenugreek plants in the future, where the ratio Gal./Man. from 1/1 (to *T. erata* is 1/1.6) (Reid and Meier, 1970), to be appropriate one-third or one-fourth for a wide industrial use. This includes pharmaceutical, textile, printing and painting industries and it may find applications in industries where starch, agar, tragacanth, acacia, carob, pectin or gelatine are at present used (Fazli and Hardman, 1968).

References

Allen, O.N. and Allen, E.K. (1981) *The Leguminosae*. Macmillan Co., London.

Anonymous (1961) *Yearbook of Agriculture*. U.S. Dept. of Agriculture, Fisheries and Food, Washington, USA.

Anonymous (1970) *Plantation crops. A review of production. trade. consumption and prices relating to coffee, cocoa, tea, sugar, spices, tobacco and rubber*, Commonwealth Secretariat.

Anonymous (1982) *The New Encyclopaedia Brittannica. – Micropaedia*, 15th edn, H. Hemingway Burton, Publ., Vol. IV, p. 94.

Anonymous (1987) *Kozieradka pospolita – Instruction of cultivation (Trigonella foenum-graecum L.)*, Instytut Roslin I Przetworow zielarskich, W. Poznaniu.

Anonymous (1994) *Plants and Their Constituents. Phytochemical Dictionary of the Leguminosae*, Vol.1, Cherman and Hall, London.

Anonymous (1996) Common catalogue of varieties of agricultural plant species. *Official J. of European Communities*, **39**, C 272 A, 45.

Arnon, J. (1972) *Crop Production in Dry Regions*. Vol.1, *Background and Principles*, Leonard Hill, London.

Awasthi, S.P. and Narayana, H.S. (1984) Effect of sucrose and sucrose plus boric acid spray on nodulation of *Trigonella foenum-graecum. Comp. Physiol. Ecol.*, **9**(1), 36–7.

Banyai, L. (1973) Botanical and qualitative studies on ecotypes of fenugreek (*Trigonella foenum-graecum* L.). *Agrobotanica*, **15**, 175–87.

Bhatti, M.A., Khan, M.T.J., Ahmed, B., Jamshaid, M. and Ahmad, W. (1996) Antibacterial activity of *Trigonella foenum-graecum* seeds. *Fitoterapia*, **67**(4), 372–4.

Bleasdale, J.K.A. and Nelder, J.A. (1960) Plant population and crop yield. *Nature* (London), **188**, 342.

Bunting, E.S. (1972) *Cultivation of Fenugreek and Some Existing its varieties*, Univ. of Feed. Lab., Oxford (personal communication).

Bunney, S. (1984) *The Illustrated Book of Herbs*, Octopus, London.

Burton, J.C. (1975) Nodulation and symbiotic nitrogen fixation. In C.H. Hanson (ed.), *Alfalfa Science and Technology*, Am. Soc. Agron. Inc. Publ. Madison, Wi., pp. 229–46.

Campbell, J.Mc A. and Reid, J.S.G. (1982) Galactomannan formation and guanosine 5-diphosphate-mannose: galactomannan mannosyltransferase in developing seeds of fenugreek (*Trigonella foenum-graecum* L.- *Leguminosae*). *Planta*, **155**, 105–11.

Ceres, E. (1984) *The Healing Power of Herbal Teas*, Thorsons Publ., Wellingborough, Northamptonshire.

Chopra, R.N., Badhwar, R.L. and Ghosh, S. (1965) *Poisonous Plants of India*, Vol. 1., Indian Council of Agricultural Research, New Delhi.

Cornish, M.A., Hardman, R. and Sadler, R.M. (1983) Hybridization for genetic improvement in the yield of diosgenin from fenugreek seed. *Planta Medica*, **48**, 149–52.

Dachler, M. and Pelzmann, H. (1989) *Heil- und Gewürzpflanzen, Anbau-Ernte-Aufbereitung*, Österreichischer Agrarverlag, Wien.

Dalianis, D.K. (1987) *Legumes for Forage, Seed and Hay*, Karaberopoulos Ltd., Athens (Greek).

Del' Gaudio, S. (1952) Il fieno greco, foraggera del colle et del monte. *Ital. Agric*, **89**, 127–36.

Del' Gaudio, S. (1953) Ricerche sui consumi idrici e indugini sull' autofertilita del fieno greco. *Ann. Sper. Agr.*, 7, 1273–87.

Duke, A.J. (1986) *Handbook of Legumes of World Economic Importance*, Plemus Press, New York and London.

Edison, S. (1995) Spices – research support to productivity. In N. Ravi (ed.), *The Hindu Survey of Indian Agriculture*, Kasturi & Sons Ltd., National Press, Madras, pp. 101–5.

Evans, W.C. (1989) *Trease and Evan's Pharmacognosy*, 13th edn, Balliere Tindall, London.

Fazli, F.R.Y. and Hardman, R. (1968) The spice fenugreek (*Trigonella foenum-graecum* L.). Its commercial varieties of seed as a source of diosgenin. *Trop. Sci.*, **10**, 66–78.

Fluck, H. (1988) *Medicinal Plants*, W. Foulsham & Co. Ltd., London.

Fred, E.B., Baldwin, I. L. and McCoy, E. (1932) Root nodule bacteria and leguminous plants. *Studies in Science*, Univ. Wisconsin Press, Ma., **5**, p. 343.

Girardon, P., Sauvaire, Y., Baccou, J.C. and Bessiøre, J.M. (1986) Identification of 3-hydroxy-4,5-dimethyl-2(5H)-furanone in aroma of fenugreek seeds (*Trigonella foenum-graecum* L.). *Lebensm.-Wiss. Technol*, **19**(1), 44–6.

Haefele, C., Bonfils, C. and Sauvaire, Y. (1997) Characterization of a dioxygenase from *Trigonella foenum-graecum*, involved in 4-hydroxyisoleucine biosynthesis. *Phytochemistry*, **44**(4), 563–6.

Hallsworth, E. (1958) *Nutrition of the Legumes*, Butterworths Scient. Publ., London.

Hardman, R. (1969) Pharmaceutical products from plant steroids. *Trop. Sci.*, **11**, 196–222.

Hardman, R. (1980) Fenugreek – a multi-purpose annual legume for Europe and other countries. *Cereal Unit Publication*, Royal Agricultural Show, Stoneleigh, UK.

Hardman, R. (1997) Utilization of fenugreek, F.R. Pharm. S. (personal communication).

Hardman, R. and Petropoulos, G.A. (1975) The response of *Trigonella foenum-graecum* (fenugreek) to field inoculation with *Rhizobium meliloti* 2012. *Planta Medica*, 27, 53–7.

Heeger, E.F. (1989) *Handbuch des Arznei- und Gewürzpflanzenbaues*, 2. Repr., Harri Deutsch Verlag, Frankfurt/M.

Hidvegi, M., El-Kady, A., Lòsztity, R., Bákás, F. and Simon-Sarkadi, L. (1984) Contribution to the nutritional characterization of fenugreek (*Trigonella foenum-graecum* L.). *Acta Alimentaria*, 13(4), 315–24.

Howard, M. (1987) *Traditional Folk Remedies, A Comprehensive Herbal*, Century Hutchinson Ltd., London.

Hughues, H. and Henson, E. (1957) *Crop production – Principles and Practices*, The Macmillan Company, New York.

Igolen, G. (1936) Fenugreek. *Parfums de France*, 14, 151–4.

Ivimey-Cook, R.B. (1968) *Trigonella* L. In T.G. Tutin, V.H. Heywood, N.A. Burges, D.M. Moore, D.H. Valentine, S.M. Walters, D.A. Webb (eds.), *Flora Europaea-Rosaceae to Umbelliferae*, Cambridge University Press, Cambridge, 2, 150–2.

Jorgensen, J. (1988) Experiments of alternative crops. *Ugeskrift for Jordbrug*, 133, 731–6.

Kamel, M.D. (1932) Reserve polysaccharide of the seeds of fenugreek. Its digestibility and its fat during germination. *Biochem. J.*, 26, 255–63.

Kamal, R., Yadav, R. and Sharma, G.L. (1987) Diosgenin context in fenugreek collected from different geographical regions of South India. *Indian J. Agric. Sci.*, 57(9), 674–6.

Khanna, P., Bansal, R. and Jain, S.C. (1975) Effect of various hormones on production of sapogenins and sterols in *Trigonella foenum-graecum* suspension cultures. *Indian J. Exp. Biol.*, 13(6), 582–3.

Kiselev, V.P., Kondrastenko, B.S., Savenko, B.I., Kodash, A.G., Zhitina, R.N. and Stikhin, V.A. (1980) Introduction of fenugreek in different areas of the USSR as a possible source of diosgenin. *Vorp. Lekarsv. Rastenievodstva*, 126–31.

Khosla, P., Gupta, D.D. and Nagpal, R.K. (1995) Effect of *Trigonella foenum-graecum* (fenugreek) on serum lipids in normal and diabetic rats. *Indian J. Pharmacol.*, 27, 89–93.

Leung, A. (1980) *Encyclopaedia of Common Natural Ingredients used in Food, Drugs and Cosmetics*, 1st edn, John Wiley & Sons, New York.

Lust, J.B. (1986) *The Herb Book*, Bantam Books Inc., New York.

Manniche, L. (1989) *An Ancient Egyptian Herbal*, British Museum Publ. Ltd., London.

Marques de Armeida, J. (1940) Study of improvement of *fenugreek* (*Trigonella foenum-graecum*). *Agronomia Lusitana*, 2, 307–35.

Max, B. (1992) This and That. The essential pharmacology of herbs and spices. *Trends Pharmacol. Sci.*, 13, 15–20.

Mercer, S.P. (1948) *Farm and Garden Seeds*, Crospy Lockwood and Son Ltd., London.

Mohamed, M.A. (1990) Differences in growth, seed yield and chemical constituents of fenugreek plants (*Trigonella foenum-graecum* L.) due to some agricultural treatments. *Egyptian J. of Agronomy*, 15(1–2), 117–23.

Molfino, R.H. (1947) Argentine plants producing changes in the characteristics of milk and its derivatives. *Rev. Farm.* (Buenos Aires), 89, 7–17.

Oliver–Brever, B. (1986) *Medicinal Plants in Tropical West Africa*, Cambridge, Univ. Press, London.

Orvedahl, C. (1962) *Good's World Atlas*. Rand Mc Nally, New York.

Pantanelli, E. (1950) La cultura delle foraggere nel mezzogiorno d'Italia. *G. Laterza*, 20, 1949–69, Bari.

Pareek, S.K. and Gupta, R. (1981) Effect of fertilizer application on seed yield and diosgenin content in fenugreek. *Indian J. Agric. Sci.*, 50(10), 746–9.

Paroda, R.S. and Karwasra, R.R. (1975) Prediction through genotype environment interactions in fenugreek. *Forage Res.*, 1(1), 31–9.

Parry, J.W. (1943) *The Spice Handbook*, Chemical Publ. Co., Brooklyn, New York.

Pattison, A.C. (1972) *Catalogue of Rhizobium Strains*, Rothamsted Experim. Station, England.

Perkins, P. (1962) *Good's World Atlas*, Rand Mc Nally, New York.

Petropoulos, G.A. (1973) *Agronomic, genetic and chemical studies of Trigonella foenum-graecum* L., Ph.D. Thesis, Bath University, England.

Piper, C.V. (1947) *Forage Plants and Their Cultures*, The Macmillan Company, New York.

Potterton, D. (1983) *Culpeper's Colour Herbal*, W. Foulsham, Slough, Berkshire.

Prasad, C.K., P.S. and Hiremath, P.C. (1985). Varietal screening and chemical control foot-root and damping-off caused by *Rhizoctonia solani*. *Pesticides*, 19(5), 34–6.

Rathore, P.S. and Manohar, S.S. (1989) Effect of date of sowing levels of nitrogen and phosphorous on growth and yield of fenugreek. *Madras Agric. J.*, 76(11), 647–8.

Reger, K.H. (1993) *Hildegard Medizin. Die natürlichen Kräuterrezepte und Heilverfahren der hl. Hildegard von Bingen*, W. Goldmann Verlag, München.

Reid, J.S.G. and Bewley, J.D. (1979) A dual role for the endosperm and its galactomannan reserves in the germinative physiology of fenugreek (*Trigonella foenum-graecum*), an endospermic leguminous seed. *Planta*, 147, 145–50.

Reid, J.S.G. and Meier, H. (1970) Chemotaxonomic aspects of the reserve galactomannan in leguminous seeds. *Z. Pflanzenphysiol.*, 62, 89–92.

Rizk, S.G. (1966) Atmospheric nitrogen fixation by legumes under Egyptian conditions. II. Grain legumes. *J. Microbiol. U.A.R.*, 1(1), 33–45.

Rosengarten, F. (1969) *The Book of Spices*, Livingston, Wynnewood, PA., USA.

Rouk, H.F. and Mangesha, H. (1963) *Fenugreek (Trigonella foenum-graecum L.). Its relationship, geography and economic importance*, Exper. Stat. Bull. No. 20, Imper. Ethiopian College of Agric. and Mech. Arts.

Saleh, N.A. (1996) *Breeding and cultural practices for fenugreek in Egypt*, National Research Centre, Cairo (personal communication).

Saleh, N., El-Hawary, Z., El-Shobaki, F.A., Abbassy, M. and Morcos S.R. (1977) Vitamin content of fruits and vegetables in common use in Egypt. *Z. Ernöhrungswiss.*, 16(3), 158–62.

Schauenberg, P. and Paris, F. (1990) *Guide to Medicinal Plants*, Lutterworth Press, Cambridge, UK.

Sekar, K. and Muthuswami, S. (1985) Economics of double intercropping in turmeric. *Indian Cocoa, Arecanut and Spices Journal*, 8(3), 67–9.

Serpukhova, V.I. (1934) Trudy, *Prikl. Bot. Genet. i selekcii Sen.*, 7(1), 69–106 (Russian).

Sharma, R.D., Raghuram, T.C. and Rao, V.D. (1991) Hypolipidaemic effect of fenugreek seeds as clinical study. *Phytotherapy Res.*, 5, 145–7.

Sharma, R.D., Sarkas, A., Hazra, D.K., Misra, I., Singh, J.B. and Maheshwari, B.B. (1996) Toxicological evaluation fenugreek seeds: a long term feeding experiment in diabetic patients. *Phytotherapy Research*, 10(6), 519–20.

Singh, D. and Singh, A. (1974) A green trailing mutant of *Trigonella foenum-graecum* L. (Methi). *Crop Improvement*, 1(1–2), 98–100.

Singh, S.N. and Rai, S.P. (1996). Companion cropping of autumn sugarcane and spices. *Indian Sugar*, 46 (3), 177–82.

Singhal, P.C., Gupta, R.K. and Joshi, L.D. (1982) Hypocholesterolemic effect of *Trigonella foenum-graecum* L. *Curr. Sci.*, 51(3), 136–7.

Sinskaya, E. (1961) *Flora of cultivated plants of the U.S.S.R. XIII. Perennial leguminous plants, Part I. Medic, Sweet clover, Fenugreek*, Israel Programme for Scientific Translations, Jerusalem.

Smith, A. (1982) *Selected markets for turmeric, coriander, cumin and fenugreek seed and curry powder*, Tropical Product Institute, Publication No. G165, London.

Stuart, M. (1986) *The Encyclopaedia of Herbs and Herbalism*, Orbis, London.

Subba-Rao, N.S. and Sharma, K.S.B. (1968) Pectin methylesterase activity of root exudates of legumes in relation to *Rhizobia*. *Pl. Soil*, 28(3), 407–12.

Talelis, D. (1967) *Cultivation of Legumes*, Agric. College of Athens, Athens (in Greek).

Vaitsis, Th. (1985) Creation of a new variety of fenugreek, named 'Ionia', resistant to *Sclerotinia sclerotiorum*. (Unpublished data), Fodder and Pastures Research Institute, Larissa, Greece.

Valette, G., Sauvaire, Y., Baccou, J.C. and Ribes, G. (1984) Hypocholesterolemic effect of fenugreek seeds in dogs. *Atherosclerosis (Shannon, Irel)*, **50**(1), 105–11.

William, A.R. and Thomson, M.D. (1978) *Healing Plants, A Modern Herbal*, Macmillan, London.

Yadar, H.D., Singh, S. and Kumar, V. (1996) Response of winter spices to sodic water irrigation in light textured sodic soil. *Haryana Agric. Univ. J. Res.*, **26**(1), 51–5.

Zade, V.R., Patil, V.N. and Zode, N.G. (1990) Standardization of seed testing procedure for *Trigonella foenum-graecum* and *Cyamopsis tetragonolobus. Ann. Plant Physiol.*, **4**(2), 182–5.

5   Breeding

Georgios A. Petropoulos

General

Fenugreek is grown under a wide range of soil and climatic conditions, in many countries of Europe, Asia, Africa, Australia and America. Its wide diversity makes any improvement a dynamic challenge.

For any crop species the nature of genetic variation, its reproductive behaviour, adaptation to different environments, the mode of inheritance of some morphological characters and usage have a bearing on the objectives and methods chosen for its genetic improvement. In addition to knowing about fenugreek phenology and reproductive system, breeders also need to be aware of its origin, existing genetic variability in the species and its wild relatives.

Fenugreek is botanically a short living (4–7 months) annual crop. Sinskaya (1961), based on the growing period, morphological characters and habits, classified fenugreek into series, subseries and ecotypes and into five groups: very early (80–85) days, early (80–90) days, mid-early, late (90–100/115) days and very late (120–140) days. Serpukhova (1934) classified the fenugreek seeds according to their shape, size and colour and distinguished three groups with six varieties in the case of one of them, while Furry (1950) also divides fenugreek according to seeds into six types (races) with names of their main habits. Serpukhova (1934) on the basis of N.I. Vavilev's collection of fenugreek in Yemen and Abyssinia, divided fenugreek into two subspecies, *iemensis* and *culta*, according to their morphological characters and the vegetation period.

As we shall see in the section on selfing and crossing, the plant is self-pollinated, but there are opportunities for natural out crossing. The inherent variation in fenugreek is quite immense and so it is grown today in the wide range of climatic conditions of all continents.

Fenugreek according to Darlington and Wylie (1945) has $2n = 16$ chromosomes, while Joshi and Raghuvanshi (1968) have investigated the presence of B-chromosomes. Singh and Singh (1976) isolated five double trisomics along with primary trisomics from the progenies of autotriploids, which had $2n + 1 + 1 = 18$ chromosomes.

The diploid nature of the normal fenugreek genetic structure is a guarantee of simplicity and existing relative experience, as diploid genetics has been evaluated extensively. So, an impressive body of information has accumulated on the theory of segregation inbreeding, selection and the genetic variances of diploids. This information is the genetic foundation for the breeding theory of fenugreek, in such a manner that the practical application of its breeding succeeds. Breeders have produced a large number of varieties and mutants that are characterised by productivity, vigorous growth, chemical and structural composition. The demand for fenugreek varieties, mainly with a higher diosgenin content, prompted more directed breeding efforts.

Origin

Sinskaya (1961) reports that the direct wild ancestor of cultivated fenugreek belonging to the species *Trigonella foenum-graecum* L. has not been exactly determined, and the existence of these wild forms (that have not escaped from cultivation) is problematic. Many authors maintain that the direct ancestor of cultivated fenugreek is the wild *T. gladiata* Ste. that differs from *T. foenum-graecum* in respect of the entire aggregate of characters, of which seed tuberculation and the small size of the pods are only the most striking. It is possible that the species *T. foenum-graecum* evolved from *T. gladiata*, which had possibly given rise to some new extinct forms of *T. foenum-graecum*.

Fenugreek is an ancient crop plant. De Candolle (1964) and Fazli and Hardman (1968) notice that fenugreek grows wild in Punjab and Kashmir, in the deserts of Mesopotamia and Persia, in Asia Minor and in some countries in Southern Europe such as Greece, Italy and Spain. De Candolle (1964) believes that the origin of fenugreek should be Asia rather than Southern Europe, because if a plant of fenugreek nature was indigenous in Southern Europe it would be far more common and not be missing in the insular floras of Sicily, Ischia and the Balearic Isles.

Serpukhova (1934) and some other authors do not fail to note that the species has probably escaped from cultivation.

Selfing and crossing

Self-pollination requires tripping the flower without introducing foreign pollen, while cross-pollination is the transmission of foreign pollen on the stigma. It is well known that for flowering plants, like fenugreek, the relative length of the stamens and pistil, the time of the anther maturing, the time of the pollen's ripening, the possibility of tripping by insects and other environmental factors such as wind, rain, heat and cold and the presence or absence of self-incompatibility and self-sterility or male sterility are the chief factors that determine what is going to occur, and give rise to self- or cross-pollination and fertilization. Allard (1960) and Darlington and Wylie (1945) have classified the plants as self- and cross-pollinated, while Del' Gaudio (1952; 1953) has studied the physiology of the fenugreek flower and has investigated its self-fertility.

The conclusion drawn from our relative experiments, observations and experience about the selfing and crossing of fenugreek can be summarised as follows:

1   After the half part of the second stage of the cleistogamous flower development, the pistil is shorter than the stamens. This to a considerable extent enables the free deposition of pollen on the stigma inside the flower (Figure 5.1) and as there is no reported phenomena of self-incompatibility, self-sterility or male sterility, self-pollination and fertilization takes place.

2   The closed form of the flower, especially of the keel, is a natural obstacle to insects from reaching the stigma of the cleistogamy type of fenugreek flower. However, if during some openings of the standard and wings, which normally occur some hours daily, an insect depresses the keel then the stamens and the stigma are made to protrude. But since the stigma is shorter than the stamens it touches the already opened anthers and the stigma is dusted by their pollen, before the lower surface of the insect touches the stigma, and still self-pollination occurs. Cross-pollination can take place only if the last fact takes place at the beginning of the second stage of flower development when either the stamens are lower than the stigma or the anthers are still closed, while the stigma is receptive and at this time cross-pollination could take place.

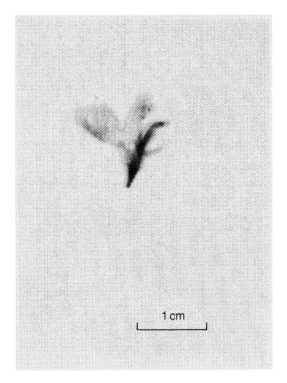

1 cm

Figure 5.1 The lower position of the pistil in comparison to the stamens, after the half part of the second stage of a cleistogamous flower of fenugreek, that enables the free deposition of pollen on the stigma, favouring self-pollination.

3 Visits by insects help self-pollination, because of the pressure on the keel and of course on the anthers, which are in touch with the keel to complete the deposition of the pollen on the stigma, inside the flower.

4 We believe that the main opportunities for natural cross-pollination of fenugreek are first via the 'aneictogamy' (open) type of flowers, especially when these are in the early stages of their development, and second the exception of the above (2) case.

5 The fraction of fenugreek cross-pollination has been estimated at 0.27 per cent (Petropoulos, 1973), but as the number of experimental plants were comparatively small more research is needed to confirm this fraction. Woodworth (1922) by alternating plants of different varieties of soybean concluded that the corresponding fraction for the soybean plant was 0.16 per cent.

6 According to Allard (1960) legumes are described as cross-pollinated crops, when frequently more than 10 per cent are out-crossed. On this basis fenugreek could be described as a rarely cross-pollinated plant.

7 More study is needed to fully understand the role of the 'open' flowers of fenugreek and the daily opening for some hours of the standard and corolla's wings.

Breeding objectives

The aim of a plant breeder is to develop improved varieties with increased yield and an acceptable grain quality and stability. This is the major breeding objective for fenugreek, as recently reported by Edison (1995) in countries such as India. Stability in production is sought by incorporating resistance or tolerance to biotic and abiotic stresses, although fenugreek has been reported tolerant to diseases, insects, drought, high pH, poor soils and salt (Duke, 1986).

Fenugreek is grown for multiple uses (Hardman, 1978) and breeding programmes need to be concerned with the suitability of the product according to its existing uses, such as high diosgenin content of seed for steroidal industry, high protein content for human and animal feeding, high mucilage (galactomannan) content with appropriate ratio of Gal./Man. for industrial uses and as the case may be for fixed oils, aromatic and spicy substances, as well as pharmaceutical constituents etc.

The proper maturity, no shattering of seed, high harvest index, more determinate growth habit and suitability for mechanisation, creation of genotype without the peculiar smell that taints the meat, milk and their derivatives, for its unlimited use as forage, are some of the other breeding objectives. According to Dachler and Pelzmann (1989) the criteria for the creation of an improved variety of fenugreek should be (i) resistance of pods to shattering, (ii) wide adaptability and (iii) uniform growth.

Not all of the above issues can be tackled simultaneously, some will not be pursued until advances in screening techniques are made and genetic variation studied further.

Genetic variation

Although Fenugreek grows well in temperate climates and the majority of the world cultivated with fenugreek is concentrated in certain countries (India, Morocco, Egypt, Ethiopia etc.), fenugreek is cultivated in most countries of all the continents for a variety of uses (food, spice, condiment, pot herb, dyeing, flavouring, perfume, mucilage, medicine etc.). It is obvious therefore that manual and natural selection has resulted in the development of plant and grain types that suitable for different uses, environments and cropping systems. So, the collaboration on an international basis for collection, evaluation, preservation and utilisation of the fenugreek germplasm is evident to all. The fact that fenugreek is diploid and self-pollinated are two factors that favour this purpose.

Edison (1995) realises that in India there is a lack of adequate genetic variability with the existing varieties and cultivars. For this reason he suggests the import/exchange of valuable germplasm, as well as promising varieties from the Mediterranean region to overcome the yield barrier and also for the production and distribution of quality planting material.

Breeding methods

In actual practice three methods namely selection, hybridisation and mutation used separately or in combination, may be involved in the development of an improved variety of fenugreek.

Selection

Selection is a simple but very important method of improving plants, as it is a basic process in plant breeding. This consists of selecting the outstanding types and discarding those that are undesirable because of certain characteristics. This method is more suitable for the improvement of fenugreek, which possesses a diploid genetic structure, and as Busbice *et al.* (1975) concluded under comparable assumptions the response to selection would be more rapid in diploid populations. Marques de Almeida (1940) investigated a new selection method for the isolation of genotypes of fenugreek, which are alkaloid-poor.

Improvement by selection method is not possible, unless the qualities of the superior types of plants can be readily detected and as it is known, differences in appearance between plants are often small and hard to detect. So, keen observation based on experience and scientific knowledge are necessary in selecting the most desirable ones.

The investigation of suitable morphological and physiological characters as an index of selection for different inherited traits of fenugreek should provide a reliable basis to predict the performance of their progenies, and could simplify relative breeding programmes. So, knowledge about the way of inheritance of ten morphological characters of fenugreek, which are presented in Table 5.1, was studied among the F2 generation plants of three crosses in order to find indexes of selection (Petropoulos, 1973) for some specific traits.

The first six characters of Table 5.1 appear to be inherited together, thus the allelomorphic genes that control them should be in the same pair of chromosomes and have a linkage. The fact that there are separate genes for these six characters has been confirmed in the field among the plants of different cultivars and populations. The linkage phenotype with the presence of the above characters has been called 'colorata', and it is completely dominant to the phenotype without these characters that has been called 'pallida' and with segregation ratios of 3:1 (Petropoulos, 1973; Cornish *et al.*, 1983). Four of the above characters are presented in Figure 5.2. The phenotype with solitary pods is also completely dominant to the phenotype with twin pods, while the phenotypes with narrow pods, large seeds and rectangular seeds are completely dominant to the corresponding phenotypes with wide pods, small seeds and round seeds.

Two procedures are commonly used for the process of selection to develop improved varieties of fenugreek: the individual or simple plant selection and the mass selection.

Individual or simple plant selection

This procedure also called pedigree and pure line selection is more effective in the case of self-pollinated plants like fenugreek, for which there is no evidence of inbreeding depression. When this method is applied individual plants of fenugreek, which like all self-pollinated species are considered normally homogynous and therefore have been selected as superior for certain characteristics according to the breeding objective and the seed from each plant, are planted in a head row of its own to give a progeny. Comparisons between the different progenies are made and those with undesirable characteristics are discarded. The superior plants are planted in longer rod rows and for fenugreek three of them are usually sown. Their plants are carefully observed

Table 5.1 List of ten morphological characters of fenugreek, for which the way of inheritance has been investigated

No.	Morphological characters		Remarks
1	Pigment in the seed coat		
2	Bluish spots on the standard		Presence (**Colorata**-type)
3	Bluish spots on the keel		
4	Bluish spots on the calyx tube	Linkage	
5	Bluish spots on the stipules		Absence (**Pallida**-type)
6	Anthocyanin on stem and leaves		
7	Number of pods/node near the top		Solitary or twin
8	Pod width		Narrow or wide
9	Seed size		Small or large
10	Seed shape		Rectangular or round

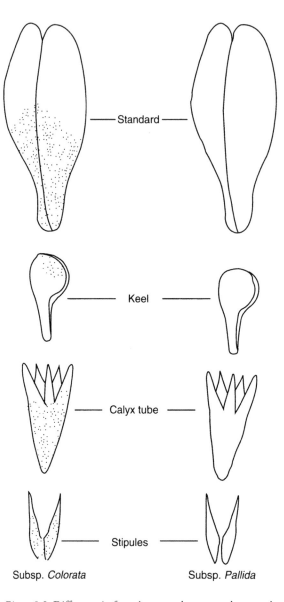

Subsp. *Colorata*   Subsp. *Pallida*

Figure 5.2 Difference in four characters between *colorata* and *pallida* type plants of fenugreek.

and the middle row is then harvested and each year the yield compared with those of the standard varieties, which are grown under the same conditions. A multiplication of the seed follows if the strain proves to be superior to the standard varieties, for distribution to farmers. For fenugreek this process takes about 8–10 years. Green *et al.* (1981) observed that pedigree selection has generally been useful in breeding for highly inheritable traits such as seed size, seed colour, growth habit and seed number per pod.

A great number of improved varieties of fenugreek have arisen by utilising this method (Del' Gaudio, 1953; Saleh, 1996).

Mass selection

Although this method is more suitable and applied to largely cross-pollinated plants, it can also be applied for the genetic improvement of fenugreek.

This method consists of selecting a fairly large number of individual plants that possess the desired characteristics to be planted in a row of its own, as in individual plant selection their seeds are mixed and bulked and sown together. The better individuals are again selected or the poorer ones discarded at anytime during this procedure. This process of selection is repeated for a few years until the plants are reasonably uniform in desired characters, according to the breeding objective. This method has helped to develop varieties of fenugreek with a higher content of diosgenin and other characteristics (Petropoulos, 1973; Saleh, 1996).

Hybridisation

Hybridisation is the crossing of two or more varieties of fenugreek that differ in one or more characteristics, which differ markedly from the parental plants in order to produce a *hybrid*. Various special techniques are used during hybridisation such as chromosome transfer by aneuploids, chromosome addition and substitution and gene transfer by translocation induced by mutagenesis.

Hybridisation offers high probability for increasing variability for further selection and the greatest possibilities for improvement of fenugreek. The parents for hybridisation should be chosen to comply with the breeding objectives and the special attributes of the lines and generally to provide planned genetic variability for subsequent selection. A special technique consisting of dialled or line tester mating schemes should be used to determine the combining ability of the varieties, that are going to be crossed.

Fenugreek is not inbred before it is crossed as it is naturally inbreeding. In fenugreek, crosses are normally made by hand, emasculation and pollination. It was found that the emasculation of its flower should be done at the end of the first stage of its development (see Chapter 4) in order to avoid selfing completely, especially for critical genetic studies. In this stage the stigma of the pistil is beginning to be receptive while the anthers of the stamens are closed and lower than the stigma. A technique of fenugreek flower emasculation is given by Cornish *et al.* (1983): after the pollination is made a bag is placed over the flower to eliminate the chance of uncontrolled cross-pollination. Successful hybridisation is generally influenced by weather conditions, particularly temperature, humidity and sunshine. After a cross has been made the progenies are grown in special plantings and the process of selection and testing are applied.

Hybridisation is a complex and time consuming process and usually hundreds of crosses must be made before an individual is found that possesses the combination of characteristics desired, but it is a method commonly used in the genetic improvement of all important seed-bearing plants and of course for fenugreek, too. By this method a great number of improved fenugreek varieties have been developed. (Petropoulos, 1973; Cornish *et al.*, 1983; Edison, 1995; Saleh, 1996).

Mutation

Plant breeding is a controlled evolution and mutation is one of the three major factors, the other two are selection and recombination. The mutation technique can be used more often in conjunction with the other breeding methods.

The result or offspring of a mutation that is called *mutant* can be utilised in various ways in plant breeding and in the case of self-pollinated species, like fenugreek, is used either immediately as a mutant, offering the greatest advantages, or in cross breeding.

Spontaneous and induced mutation can be distinguished. Spontaneous mutation is produced mainly by cosmic radiation and is the main factor of natural evolution ultimately responsible for all variability in living things. Several mutants of fenugreek from spontaneous mutations have been isolated and today are in use all over the world (Petropoulos, 1973; Singh and Singh, 1974; Laxmi *et al.*, 1980; Laxmi and Datta, 1987). The interest in induction of mutations in plant breeding has increased considerably all over the world in the last thirty years (Dubinin, 1961; Anonymous, 1961; Manha *et al.*, 1994).

For diploid plants, like fenugreek, the majority of the induced mutations are recessive and segregate in a 3 : 1 ratio (Gaul, 1961; Petropoulos, 1973; Singh and Singh, 1974). For the induction of mutations in fenugreek and other plants, ionising radiations and different chemical mutagens are used, while genetic engineering, which is a recent biotechnological speculation, concerning the potential impact of new technique in cell and molecular biology on plant improvement, could also be used.

Ionising irradiation, which includes electromagnetic radiation (x- and gamma rays) and the so-called particulate radiation (alpha, beta, protons, etc.), is used to artificially increase the rate of spontaneous mutations.

The effects of ionising radiations are on nuclei, chromosome (breakage and aberration) and genes. In practical breeding work, selection of mutants can start in the M1 generation, but are commonly done in the M2 generation and should be continued in the M3 and following generations.

In an effort to induce mutations in fenugreek two methods were applied: gamma-irradiation of isotope Cobalt-60 as chronic rays in an open irradiated field, and acute rays on the dry seeds (Petropoulos, 1973). A 26 m diameter open field at Bath University with a source of Cobalt-60 in the centre, which held 165 millirads/hour, was used. A special mechanism was operated to raise the source (see Figure 5.3) or to lower it into its protective lead shield (see Figure 5.4) during visits to the experimental area. The seeds were sown in twelve orbit rows each being at a distance of 1 m apart (see Figure 5.5). The amount of irradiation received by fenugreek plants is presented in Figure 5.6, while the corresponding irradiation received by its reproductive organs is presented in Figure 5.7. Although the source proved quite low, interest was concentrated on the seeds of the first and less on those of the second row plants, where some promising mutants, which are described in the following sections, were isolated.

Acute gamma ray application on the dry seeds was used to investigate the relative sensitivity of fenugreek. For four cultivars the 'critical dose' in which about 40 per cent of the plants survive was found to be: Fluorescent 140–145 Kr, Ethiopian 135–140 Kr, Kenyan 110–120 Kr and Moroccan 140–145 Kr. A delay in flowering (Figure 5.8) and a decrease of height (Figures 5.9 and 5.10) and seed yield (Figure 5.11) were found. The main reason for the marked depression in plant growth appeared to be the reduction in root length (see Figure 5.12). It was found that, in order to produce useful fenugreek mutants, the applied dose should be much lower (50–60 Kr), and some promising mutants, which are described in the following sections, are isolated (Petropoulos, 1973).

The chemical mutagens belong to different groups and very little is known about the action of most of them (Auerbach, 1961). A lot of fenugreek mutants have been isolated by the treatment of dry seeds with different chemical mutagens (Laxmi *et al.*, 1980; Singh and Raghuvanshi, 1980; Laxmi and Datta, 1987; Jain and Agrawal, 1987), while shoot apexes of fenugreek treated by colchicine produced tetraploid plants with promising economic characteristics (Roy and

Figure 5.3 A radiation device (installation) with the special raising mechanism for irradiating the source, in operation.

Figure 5.4 The same device with the special mechanism to lower the source into its protective lead shield, during visits to the experimental area.

Figure 5.5 Orbitic sowing of the field irradiation area 1 m apart.

Singh, 1968). The effect of mutagens on tissue cultures of fenugreek with UV-irradiation and methyl methane sulphonate, increased steroidal sapogenin about two- to three-fold (Jain and Agrawal, 1994).

Breeding for higher yield

Increased yield with acceptable seed quality and stability is determined by a complex interaction between its genetic makeup and environmental (biotic and abiotic) factors. In the genetic factors are included the production of a higher number of pods with more and larger seeds, the proper precocity, the uniform maturity, the resistance to lodging, the less shattering of pods and scattering of seeds, etc. Biotic stresses include diseases, pests and weeds, while abiotic limitations mainly include temperature, moisture and wind.

Improved stability and performance are obtained from varieties that incorporate resistance/ or tolerance to the above stresses, although fenugreek generally tolerant to most of these biotic and abiotic stresses and eliminations (Sinskaya, 1961; Fazli and Hardman, 1968; Duke, 1986).

Traditional yields have been measured in terms of dried seeds (11 per cent moisture) per acre and per year. As Edison (1995) reports, in India, among the efforts to increase productivity of spices is the evolution of high yielding varieties with greater stability. It must be emphasized that there is a negative correlation between yield and quality, in general.

Del' Gaudio (1953) by selecting a single fenugreek plant with a short winged flower, created the very productive cultivar 'Ali Corte.'

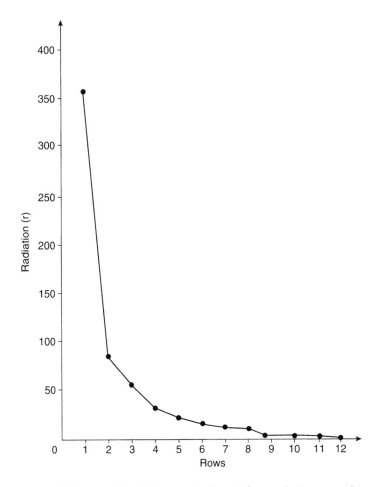

Figure 5.6 Amount of irradiation received by the fenugreek plants according to their distance from the center of the source.

Using the rich collection of fenugreek at the University of Bath, England, (Petropoulos, 1973) the following productive and promising hybrid and mutant genotypes have been isolated:

1   The genotypes RH 3109/32, RH 3110/37, RH 3105/15 and 3111/8, which are products of crosses between the cultivars Fluorescent and Kenyan, gave seed yields of more than two-fold over the average of those from the best parents.
2   The mutant RH 3112 from induced mutation in the open field of irradiation from the Kenyan cultivar, gave a seed yield of almost double that of the mother cultivar.
3   The selected line RH 3128 from the Kenyan cultivar gave a seed yield that was more than double of the corresponding yield from the mother cultivar.

Breeding for a superior quality of yield

General

The contribution of plant breeding to the creation of improved varieties of superior quality seed is well documented. The quality of fenugreek seed affects its value, ultimate use, how it is

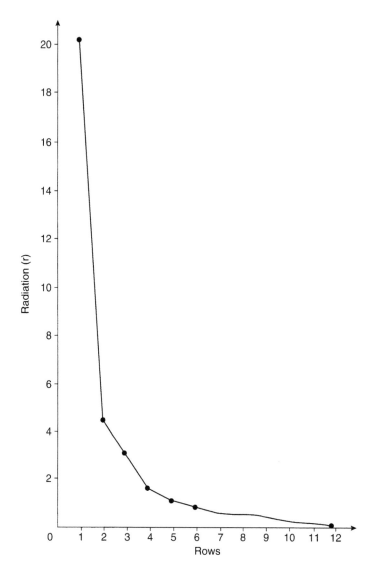

Figure 5.7 Amount of irradiation received by the reproductive organs of the fenugreek plants, according to their distance from the center of the source.

processed and further affects the humans or livestock consuming the seed. The term quality is difficult to define because the grower, the processor and the ultimate user have different criteria for determining quality. The quality of seed definition in establishing grades and prices is based on colour of seeds, freedom from diseases and pests, low percentage of shrivellent seeds and high percentage content of the active constituents (diosgenin, protein, mucilage, medicinal, spicy, etc.). This is without estimating the influence of post harvest treatments like extraneous matter and impurities, proper moisture content, etc. Breeders need better definitions of inheritable characteristics contributing to quality, as well as better techniques for their measurements in segregating populations. The quality obtained in the mature fenugreek seed is a result of both the genetic make-up of the plant and the environment in which it grew.

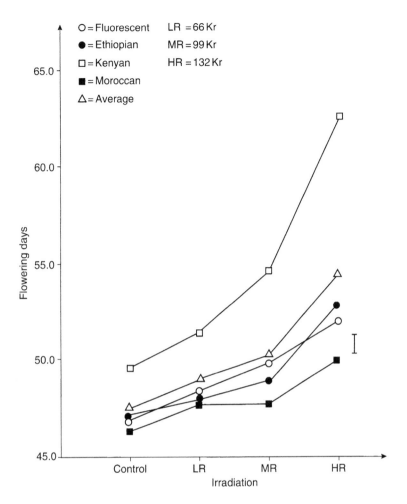

Figure 5.8 Correlation between seed irradiation dose with acute gamma rays and flowering days of fenugreek.

The environmental influences can be divided into physical (temperature, wind, precipitation, soil fertility) and biological (diseases, pests and weeds).

Genetic effects for the improvement of seed quality usually determine importance and effectiveness, because many quality traits of fenugreek are inheritable. These traits are either simple and controlled by a small number of genes or inherited more complexly, and are difficult to manipulate in a breeding programme (Collins and Petolino, 1984). The genotype × environment interaction is usually the reason of the failure of a genotype to perform similarly in different environments. However the final goal of a fenugreek breeder is the development of a variety of excellent quality over a wide range of environments (Paroda and Karwasra, 1975).

When genetic variance for a desirable trait in a breeding population is low compared to non-genetic influence on the trait, selection procedures become more complex, often involving progeny testing in replicated trials in varying environments (Collins and Petolino, 1984).

The effort of the plant breeder to develop fenugreek varieties like other species, with superior quality seed for certain special traits could result in commercially unacceptable varieties, unless

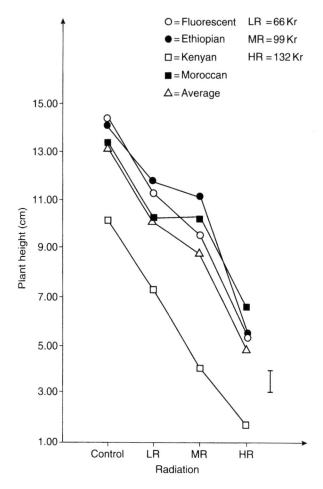

Figure 5.9 Correlation between seed irradiation dose with acute gamma rays and height of fenugreek plants.

the marketplace will pay a premium for the improved quality to compensate for the probable lack of an improved yield (Collins and Petolino, 1984).

Several notable achievements have been obtained concerning the development of improved varieties of fenugreek with superior quality seed such as higher content of diosgenin in the seed, protein, fixed oils and mucilage.

Breeding for a higher diosgenin content in the seed

The steroidal diosgenin is a monohydroxysapogenin, and it is of importance to the pharmaceutical industry as a starting material in the partial synthesis of corticosteroids, sex hormones and oral contraceptives.

Hardman (1969) considers the fenugreek seed to be of commercial interest as it is a source of diosgenin, but its content is relatively low for economical and beneficial exploitation. There are some possibilities for increasing the diosgenin contained in the seed, either during the growing

Figure 5.10 Reduction in height of fenugreek plants, due to seed irradiation with acute gamma rays.

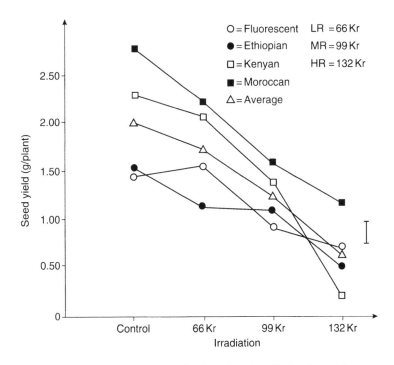

Figure 5.11 Correlation between seed yield and seed irradiation dose with acute gamma rays.

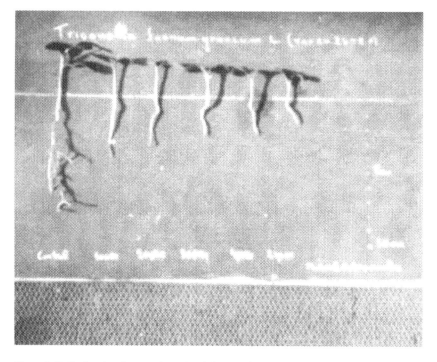

Figure 5.12 Reduction in root length of fenugreek plants, due to seed irradiation with acute gamma rays.

period by using different cultural techniques (Kozlowski *et al.*, 1982; Mohamed, 1983), or during post harvest treatments by different techniques (enzymes, hormones, etc.) of germination with incubation (Hardman and Fazli, 1972a,b), different conditions of incubation (Elujoba and Hardman, 1985a), and fermentation (Elujoba and Hardman, 1985b), by storage (Hardman and Brain, 1972), by the use of tissue and cell culture (static or suspension) (Khanna and Jain, 1973; Stevens, 1974; Stevens and Hardman, 1974; Khanna *et al.*, 1975; Hardman and Stevens, 1978; Trisonthi *et al.*, 1980) and by biological manipulation of the steroidal yield (Hardman and Brain, 1970). However, the main effort is still the increase of the diosgenin content by genetic improvement of plant. In the seed, diosgenin is present only in the embryo, but it is absent from the testa and endosperm (Fazli and Hardman, 1968), it is also in other parts of the plant (stems, leaves etc.) but their content is very low (Hardman and Fazli, 1969; Varshney *et al.*, 1980).

The demand for fenugreek varieties with a higher diosgenin content in the seed prompted more directed breeding efforts. The diosgenin content is an inheritable character (Petropoulos, 1973; Cornish *et al.*, 1983) and as a quantitative one should be controlled by more than one gene (Poehlman, 1979). There are also indications that the diosgenin content of fenugreek depends on genotypic and geographical differences (Kamal *et al.*, 1987). The F1 generation can be seen as intermediate, while the F2 shows a wide range of concentrations (Cornish *et al.*, 1983). Also, in a case of a cross there was no evidence of potency or epistasis in the control of diosgenin content and the broad inheritability was estimated at around 40 per cent. This indicates a significant segregation of the genes controlling the diosgenin yield (Cornish *et al.*, 1983). Sufficient genetic variation exists in the yield of diosgenin from fenugreek seed that permits a plant breeder using a suitable breeding method, to select promising lines for increased diosgenin content (Cornish *et al.*, 1983).

The diosgenin content of the fenugreek seed according to Duke (1986) fluctuated between wide limits ranging from 1–2 per cent. Sharma and Kamal (1982) reported the diosgenin content of the seed as 0.33–1.90 per cent from seeds collected from different regions of India. It must be emphasized that the results about the diosgenin content of fenugreek seed, which are presented in the related literature are not comparable all the time for different reasons. First because these have been determined by different analytical methods (infrared spectrometry, combined column chromatography, thin-layer chromatography, gas and liquid chromatography, etc.) as have been described by various researchers (Hardman and Jefferies, 1972; Dawidar and Fayez, 1972; Jefferies and Hardman, 1972; Dixit and Srivastava, 1977) and second these express dissimilar things, that is, pure diosgenin or the natural mixture of diosgenin plus yamogenin in the ratio 3:2 as commonly found in fenugreek seed (Cornish *et al.*, 1983). Pasich *et al.* (1983) have reported this ratio to be 2:1.

The development of improved varieties of fenugreek with a higher diosgenin content in the seed should be obtained at first from the existing populations, cultivated or landraces, using known breeding methods especially those with the induction of mutations.

Apart from better definitions of inheritable characteristics contributing to the higher diosgenin content, plant breeders need accurate and mainly quicker techniques for their measurements in segregating populations. Since the determination of diosgenin content is, at the moment laborious, the evaluation of a large number of isolated progenies is very difficult. So, the investigation of a selection index for a higher diosgenin content of seed based on the morphological characters of the fenugreek plant for a rough detection and isolation of the most promising progenies, is desirable. There are firm indications that there is a linkage of the quantitative character of diosgenin content with the morphological character of the number of pods per node *near the top of stem*, and that high content of diosgenin is inherited together with the formation of twin pods. So, the phenotype of twin pods in comparison with that of solidary pods is a good index of selection and should provide a reliable basis to predict the performance of their progenies for a higher diosgenin content of seed, from very early generations. This will simplify planned research programmes of genetic improvement for this purpose (Petropoulos, 1988). The index of twin pods could be utilised without any decrease of seed yield, as there is no correlation between the property of twin pods and seed yield. The superiority of this phenotype was confirmed when it was used as a criterion of mass selection in the case of the creation of Moroccan and Kenyan cultivars, where the diosgenin content was increased by 23 and 12 per cent respectively (Petropoulos, 1973).

A lot of improved fenugreek varieties, cultivars and promising genotypes, as far as higher diosgenin content of seed is concerned, have been developed through the utilisation, as the case may be, of one or more of the known breeding methods (Cornish *et al.*, 1983). From the breeding work at Bath University, England, using a rich collection of fenugreek the following improved cultivars and promising hybrid progenies and mutants have evolved (Petropoulos, 1973).

1   The cultivar Moroccan (RH 2701) with 1.19 per cent diosgenin, that was created by continuous mass selection of the population RH 2283, which originated from Morocco, with 0.97 per cent diosgenin (progress 23 per cent).
2   The cultivar Fluorescent (RH 2602) with 1.38 per cent diosgenin that evolved by spontaneous mutation from the Ethiopian population RH 2475 with 1.18 per cent diosgenin (progress 17 per cent).

3 The cultivar Kenyan (RH 2698) with 1.51 per cent diosgenin that was created by continuous mass selection of the population RH 2591, which originated from Kenya, with 1.35 per cent diosgenin (progress 12 per cent).

4 The hybrid progeny RH 3109/42 from a cross of Fluorscent × Kenyan, with 1.83 per cent diosgenin (progress 21 per cent over the best parent).

5 The hybrid progeny RH 3110/66 from another cross of Fluorscent × Kenyan with 1.81 per cent diosgenin (progress 20 per cent over the best parent).

6 The mutant RH 3112 induced by gamma rays in an open field irradiation, with 1.78 per cent diosgenin from the Kenyan cultivar with 1.51 per cent diosgenin (progress 15 per cent over the mother cultivar).

7 The spontaneous mutant RH 3129 with a shorter and broader standard of flower and high proportion of twin pods, with 1.35 per cent diosgenin from the Moroccan cultivar with 1.19 per cent diosgenin (progress 13 per cent).

It must be emphasized that in all of the above cases the main criterion of selection or detection was the high proportion of twin pods of the plants, near the top of the stem.

Breeding for higher protein content of seed

The fenugreek seed is quite rich in protein content in comparison with other cereal grains and legumes (Petropoulos, 1973; Awadala *et al.*, 1980; Ullah, 1982), but the increasing protein deficiency all over the world, justifies every effort for the genetic improvement of fenugreek in this direction. This will also help in the easier valorisation of the by-products for animal feeding, after the probable extraction of diosgenin for industrial purposes.

The genetic variability for protein content among a collection of 123 hybrid lines of fenugreek that was varied from 20.4–39.3 per cent have been reported (Petropoulos, 1973), while Duke (1986) gives an average of 23.2 per cent. Hidvegi *et al.* (1984) reported a protein content of 26.4 per cent for their samples.

The protein quality of fenugreek seed, calculated from the amino acid pattern in comparison with the data for human requirements, approaches that of the soybean (Hidvegi *et al.*, 1984). The same researchers report that fenugreek protein is rich in lysine, higher than that found in an 'average legume', but it has a relatively low (32 per cent) multienzymatic digestibility and bitter and anti-nutritive components, mainly because of the sapogenin content. Duke (1986) reports also that fenugreek protein is rich in lysine, but poor in S-amino acids and tryptophan. The same author gives the analytical composition of fourteen amino acid values for fenugreek protein (percentage of protein).

To increase the crude protein content of fenugreek seed, the relationship of several morphological and physiological characters to protein content was investigated (Petropoulos, 1973; 1990), and such relationships for different traits are used for many plants (Olson, 1960; Evans, 1984; Tungland *et al.*, 1987). Evaluation of the phenotypic correlation of a lot of characters of F2 plants, of three crosses, indicated that among these characters four of them namely wide pods, fluorescent under UV light seeds, large seeds, and ellipsoid (round) in outline seeds, were proved superior to the corresponding opposite phenotypes: narrow pods, no fluorescent seeds, small seeds and rectangular in outline seeds, as far as the protein content of seed is concerned. Regression analysis of this data showed that the simultaneous presence of these four favourable phenotypic characteristics in the same plant gives the best results for protein content (Petropoulos, 1990). In these four favourable phenotypes only the large seeds are controlled by dominant incomplete genes, while the other three are controlled by recessive genes and so bred true. Also the two phenotypes: fluorescent seeds and wide pods are essential as far as the protein content is concerned, as they represent 74 per cent of the total variability for protein, bred true and are easily detected in the field (Petropoulos, 1990). The superiority of the favourable pheno-types was confirmed when they were used as the criteria of mass selection in three fenugreek populations, where the protein content has been increased by a positive selection from 11–17 per cent and decreased by negative selection (use of opposite phenotypes) by 10 per cent. A sim-ple explanation of this superiority is that these favourable phenotypes correlate with a higher g/h index of seeds and it is well known that the protein content of the germ is higher than that of the husk.

The correlation between the protein content of fenugreek seed and the number of favourable phenotypic characters in the same plant is linear and follows the equation: $y = 23.94 + 2.208\,x\,(r = 0.9092)$ (Figure 5.13).

The use of the favourable phenotypes and mainly the simultaneous presence in the same plant of more than one as an index of selection, provides a reliable basis to predict the performance of their progenies for higher protein content from very early generations. This will simplify planned research programmes for genetic improvement in this direction without any decrease of seed yield, as there is no indication of any relationship between seed yield and protein content in fenu-greek (Petropoulos, 1973; 1990) in contrast with other plants, where a negative correlation has been found (Caldwell *et al.*, 1966). For example in soybean increased grain protein percentage is commonly associated with reduced grain yield per unit of land area.

The above observations apply on the condition that these results are tested in one and only one environment for it is known that these phenotypes and the protein contained in other plants are influenced by environmental conditions (Ries and Everson, 1973). A lot of improved fenugreek varieties, cultivars and promising genotypes, regarding higher protein content of seed have been developed using different breeding methods.

At the University of Bath following extensive breeding work using a rich collection of fenu-greek samples from all over the world and applying known breeding methods, the following improved cultivars and promising lines, as far as higher protein content of seed is concerned, were developed (Petropoulos, 1973):

1   the cultivar Ethiopian (RH 2699) with 32.95 per cent protein, by continuous mass selection of the population RH 2278, originated from Ethiopia, with 28.06 per cent protein (progress 17 per cent);

2   the hybrid progeny RH 3109/39 from a cross of Fluorescent × Kenyan with 39.29 per cent protein (progress 27 per cent over the best parent);

3   the hybrid progeny RH 3110/83 from a cross of Fluorescent × Kenyan with 36.31 per cent protein (progress 18 per cent over the best parent);

4   the hybrid progeny RH 3109/37 from a cross of Fluorescent × Kenyan with 35.56 per cent protein (progress 15 per cent over the best parent);

5   the mutant RH 3112 with 32.29 per cent protein, from induced mutation with chronic gamma rays in an open irradiation field, from the Kenyan cultivar (progress 34 per cent);

6   the mutant RH 3118 with 28.42 per cent protein, from induced mutation, also from Kenyan cultivar (progress 15 per cent);

7   the mutant RH 3115 with 30.56 per cent protein from induced mutation using acute gamma rays on dry seeds from Kenyan cultivar (progress 26 per cent).

Breeding for higher fixed oils content of seed

Fenugreek seeds contain *c.* 8 per cent oil (Petropoulos, 1973; Duke 1986) extracted by either, but Varshney *et al.* (1980) report an oil content of 20 per cent. It possesses a strong celery odour

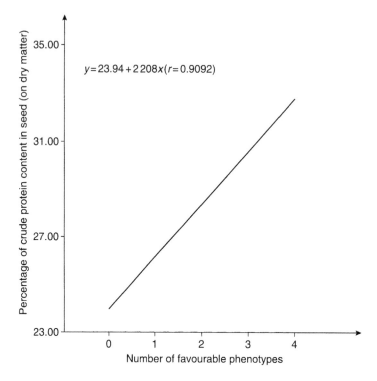

Figure 5.13 Correlation between protein content of fenugreek seed and the number of favourable phenotypes of plant to this direction.

and is used in butterscotch, cheese, licorise, pickle, rum, syrup and vanilla flavours and may be of interest to the perfume industry (Duke, 1986).

Oil content as quantitative character is a trait in which, as in soybean and other plants, a number of genes are involved (Brim, 1973). Such characteristics are quantitatively inherited by considerable environmental influence (Collins and Petolino, 1984).

The proportion of fatty acids in the oil affect oil quality. In fenugreek fixed oils the proportion of fatty acids to percentage of total acids is: 35.1 per cent oleic, 13.8 per cent linolenic, 9.6 per cent palmitic, 4.9 per cent stearic, 2.0 per cent arachidic, 0.9 per cent behenic, 33.7 per cent linoleic (Duke, 1986), while Varshney *et al.* (1980) report sapon values 202, unsaponifiable matter 0.9 per cent and 30 per cent octadecatrienoic acid. For the fatty acids composition of many plants, several simply inherited alleles that alter this composition, have been reported (Collins and Petolino, 1984). The breeding effort for fenugreek should be like that of soybean selection for high oleic acid and low linolenic acid, as this improves the flavour and stability of the oil (Collins and Petolino, 1984). The linolenic acid content, as in soybean, exhibited significant genotype × environment interaction, therefore the selection for low linolenic acid content should be done over locations and years (Collins and Petolino, 1984).

Most seed quality characteristics are related to the relative chemical and physical characteristics of the seed and as such many of them are interrelated. For example, in fenugreek and especially in the cultivar Moroccan (RH 2701) an increase of oil content by 13 per cent resulted in a corresponding decrease of protein content by 11 per cent, in comparison with the original population RH 2283, indicating a negative correlation between the oil and protein content in fenugreek seed.

Although the breeding effort to increase the fixed oils content in fenugreek, as in most grain crops, is quite difficult as this characteristic tends to be under multigenic control, some interesting cultivars have been developed in this direction:

1   The cultivar Moroccan (RH 2701) with 8.10 per cent fixed oils created by continuous mass selection, was proved superior in fixed oil content by 13 per cent over the original population RH 2283, with 7.14 per cent oil (Petropoulos, 1973).
2   Among mutant plants of fenugreek after induced mutations by chemical mutagens and gamma rays, some of them were detected for their superiority in oil content over the control (Laxmi *et al.*, 1980).

Breeding for higher mucilage (content and quality) of seed

The well developed endosperm of the fenugreek is rich in polysaccharide mucilage (galactomannan) that possesses high viscosity and neutral ionic properties (Duke, 1986), which can be used widely in industry including pharmaceutical, cosmetics, hair preparations, paper products, paints and plasters. The industrial use of fenugreek galactomannan is limited because of its inappropriate ratio of Gal./Man. which is around 1 : 1, while the appropriate ratio is 1 : 3 or 1 : 4 (Reid and Meier, 1970).

The main breeding effort should be toward the creation of improved fenugreek varieties with the appropriate Gal./Man. ratio, using a special methodology in which genetic engineering techniques may be included. The identification of the mechanisms of fenugreek galactomannan biosynthesis during seed development, and galactomannan hydrolysis during germination, may be of help in this direction.

The genetic variability of fenugreek for mucilage was found to contain 17–22 per cent (Petropoulos, 1973), while Duke (1986) reports that fenugreek seeds contain 26.3 per cent mucilaginous material. The effort of a plant breeder to develop fenugreek varieties with higher mucilage content could result in commercially unacceptable varieties, as an increase in such characteristics may lead to a reduction of some others, unless the marketplace will pay a premium for the increased mucilage content to counterbalance the probable reduction of some other characteristics.

It was found that some samples of Indian origin possessed a higher mucilage content of seeds followed by samples of Ethiopia, while samples from the Mediterranean area were inferior as far as the mucilage content is concerned (Petropoulos, 1973).

Breeding for resistance to diseases

General

Even though fenugreek is considered to be a disease tolerant crop (Sinskaya, 1961; Duke, 1986), it often suffers from various diseases, especially under environmental conditions favourable for the development of corresponding fungi and viruses (Sinskaya, 1961; Gopal and Maggon, 1971; Raian *et al.*, 1991). The *Review of Applied Mycology* (Anonymous, 1968) lists a number of fungi infecting fenugreek.

Plant resistance is the most practical means of controlling most fenugreek diseases, and so the development of varieties resistant to economically important diseases is a big contribution from fenugreek breeding programme. The goal of such a fenugreek programme should be the incorporation of resistance to as many important diseases and insects as possible, without disturbing desirable agronomic traits. Varieties of fenugreek which are resistant to disease provide built-in insurance for growers, at a very low cost.

The principles of breeding for disease resistance of fenugreek are those used for breeding most of its other characteristics, except that knowledge of the pathogen, the host and host × pathogen interactions, is needed.

Breeding for resistance to specific diseases

The diseases and viruses for which some breeding work or evaluation for incorporation of resistance or tolerance has been carried out are as follows:

Root rot (Rhizoctonia solani). It appears to be the most important root disease of fenugreek (Prasad and Hiremath, 1985; Raian *et al.*, 1991; Haque and Ghaffar, 1992).

Prasad and Hiremath (1985) reported that among twenty varieties of fenugreek screened for their resistance against *Rhizoctonia solani* (colour rot) only TG-18 and UM-20 showed some tolerance, by giving 43.3 and 35.2 per cent seedling stand respectively, a week after sowing, while none of the varieties tested showed complete resistance.

Leaf spot (Ascochyta sp.). Leaf spot is a serious disease afflicting fenugreek (Anonymous, 1968; Petropoulos, 1973) like most other legumes (Walker, 1952; Anonymous, 1970). Inheritance of leaf spot resistance of fenugreek is not fully understood.

Selection of resistance in leaf spot-sick plots was carried out in two locations during experimentation and after continuous evaluation some leaf spot tolerant cultivars and mutants were isolated (Petropoulos, 1973). So variation in the sensitivity to attack by the fungus *Ascochyta* sp. was found among four breeding cultivars as is indicated in Table 5.2. This table shows that the cultivars Ethiopian and Fluorescent had consistently low disease levels and consequently should be considered as tolerant to this disease, while the cultivar Moroccan is the most susceptible. Also, phenotypic selection for leaf spot resistance among different induced mutants resulted in the isolation of the following resistant ones:

1   The mutant RH 3113 from induced mutation using chronic gamma rays in an open irradiation field from Moroccan cultivar.
2   The mutant RH 3118 from induced mutation, like the above, but from Kenyan cultivar.
3   The mutant 3122 from induced mutation by seed irradiation with acute gamma rays from Kenyan cultivar.

Powdery mildew (Oidiopsis sp.) Palti (1959) has described this disease in Israel and considers it one of the most important diseases of fenugreek, while Rouk and Mangesha (1963) report that in Ethiopia fenugreek is attacked by powdery mildew, which does considerable damage to the plants. Petropoulos (1973) reports attacks of this fungus in fenugreek plants in England, although Agrios (1969) reports that this fungus causes damages in arid and semi-arid environments and is not favoured by wet weather. Inheritance of the powdery mildew resistance of fenugreek is not fully understood.

Four breeding fenugreek cultivars were evaluated for their susceptibility to powdery mildew and the results are presented in Table 5.2. According to these results the Fluorescent and the Ethiopian cultivars were found fairly tolerant to the powdery mildew, while the Moroccan cultivar was proved the most susceptible and this difference appears in Figure 5.14.

Pod Spot (Heterosporium sp.). This disease was described for the first time by Petropoulos (1973). The inheritance of this new disease in fenugreek has not been studied yet.

The four breeding fenugreek cultivars were evaluated for their susceptibility to the pod spot and the relevant results are presented in Table 5.2. The Kenyan cultivar was proved tolerant and

Table 5.2 Sensitivity of four breeding cultivars of fenugreek to attacks by three different pathogens

No.	Diseases	Sensitivity of cultivars (in angles)			
		Fluorescent	Ethiopian	Kenyan	Moroccan
1	Pod spot (*Heterosporium* sp.)	5.1*	5.2	1.7	2.5
2	Leaf spot (*Ascochyta* sp.)	5.7	5.6	13.9	24.5
3	Powdery mildew (*Oidiopsis* sp.)	2.4	2.7	13.3	29.6

Note
* A high figure indicates that the cultivar shows the character to a high degree.

Figure 5.14 Susceptibility of Moroccan cultivar of fenugreek to attacks by the fungus *Oidiopsis* sp.

the Moroccan cultivar fairly tolerant, while the Fluorescent and the Ethiopian were found the most susceptible to the pod spot disease. The fact that both these susceptible cultivars are late in maturing suggests that their sensitivity is linked to their stay in the tender form for a longer time. There are indications that the inoculation of the fenugreek seed with *Rhizobium meliloti* increases the sensitivity of the plants to attack by the *Heterosporium* sp.

White mould (Sclerotinia sclerotiorum). White mould is the most important fenugreek disease in Greece. The inheritance of resistance to white mould in fenugreek is not fully understood.

Figure 5.15 Aphid and mechanical transmission of BYMV to fenugreek plants. (a) aphid transmission (mild symptoms) and (b) mechanical transmission (severe symptoms).

Vaitsis (1985) working on fenugreek breeding isolated a clone resistant to this disease, after multiplication of seed the released variety named 'Ionia' is considered resistant to this fungus (Anonymous, 1996).

Bean Yellow Mosaic Virus (BYMV). There are strong indications that a common gene controls the resistance of fenugreek to BYMV and this is supported by Schroeder and Providenti (1971) in the case of *Pisum sativum*. An experiment was carried out within the facilities of the Glasshouse Crops Research Institute at Littlehampton (Brunt, 1972) in order to investigate any resistance to BYMV infection among four breeding cultivars (Fluorescent, Ethiopian, Kenyan, Moroccan) using the techniques of aphid and mechanical inoculation (transmission) of the virus.

An interaction was found between cultivars and mode of transmission, as far as the severity of infection is concerned. So, the Fluorescent cultivar showed the mildest symptoms in the case of the aphid transmission and the most severe in the case of the mechanical one, while the Moroccan cultivar showed the opposite result.

Totally, the plants infected by mechanical transmission showed more severe symptoms (dwarfness and chlorosis) than those infected by aphid transmission (Figure 5.15). This is very favourable, as aphid transmission is the only mode of infection in the field.

The conclusion drawn from this experiment is that in the field the Fluorescent cultivar is more resistant to BYMV than the other three cultivars, followed by the Ethiopian one, while the Moroccan cultivar is the most susceptible.

Breeding for special traits

Precocity

Precocity is pursued mainly under adverse (usually wet) climatic conditions in fenugreek cultivation. Inheritance of fenugreek precocity is not fully understood. The criteria of selection for precocity used in England were the earliness of flowering, the shorter duration of the stage of pod ripening that is usually more than 25 days and the limited appearance of the property of indeterminate growth habit (Petropoulos, 1973).

Phenotypic selection and screening for higher precocity among the four breeding cultivars and mutants induced by different methods, resulted in several promising precocious ones. The variation in the precocity of the four breeding cultivars is indicated in Table 5.3.

Thus the Moroccan cultivar is the earliest of the four cultivars in the ripening of pods, followed by the Kenyan and Ethiopian ones, while Fluorescent is the last to achieve maturity (at least 20 days in comparison with the Moroccan).

Also phenotypic selection and screening among mutant plants induced by chronic gamma rays in an open irradiation field and dry seed irradiation by acute gamma rays was effective, and resulted in the following precocious mutants (Petropoulos, 1973):

1   The mutant RH 3112 induced by chronic gamma rays from the Kenyan cultivar that is earlier than the mother cultivar by 20 days.
2   The mutant RH 3114 induced by chronic gamma rays from the Fluorescent cultivar whose pods are ripening simultaneously with the Moroccan cultivar, which is characterised by its earliness of ripening (progress of the mutant over the mother cultivar by 14 days).
3   The mutant RH 3116 induced by chronic gamma rays from the Fluorescent cultivar also starts to ripen at the same time as the Moroccan cultivar (progress 14 days).

Singh and Singh (1974) reported the isolation of the mutant named 'Trailing Green' induced by spontaneous mutation from the clone 'IC-74' that flowers 30 days earlier than the mother clone.

Resistance to lodging

Although the fenugreek stem is naturally erect and strong, there is always the danger of strong winds and rains that cause lodging and the crop to lay down. So, the creation of fenugreek varieties resistant to lodging, especially in areas where strong winds predominate have a high priority.

Table 5.3 Precocity of four breeding cultivars of
fenugreek

No.	Cultivars	Earliness of ripening (%)
1	Fluorescent	31.3*
2	Ethiopian	37.2
3	Kenyan	42.3
4	Moroccan	53.2

Note
* The figures express the percentage average of the
pods reached in full maturity under UK conditions.

Table 5.4 Resistance to lodging of four
breeding cultivars of fenugreek

No.	Cultivars	Resistance to winds (scale 1 α 5)
1	Fluorescent	1.4*
2	Ethiopian	1.9
3	Kenyan	3.0
4	Moroccan	3.9

* A high figure indicates that the cultivar
shows the character to a high degree.

The criteria for resistance to lodging that have been applied at Bath University were: the shortness and thickness of the shoots, the presence of a shoot hollow that is as narrow as possible and the production of secondary shoots arising from the base of the stem (Petropoulos, 1973). After a 2 year evaluation, the variation in lodging resistance of the four breeding cultivars is indicated in Table 5.4. Thus the Moroccan cultivar followed by the Kenyan are the most resistant to lodging because of the short and narrow hollow shoots and the presence of secondary shoots arising from the base, while the most susceptible is the Fluorescent cultivar because of its tall and wide hollow shoots and the absence of secondary shoots from the base.

Phenotypic selection and screening for lodging resistance among mutant plants induced by irradiation with chronic gamma rays of the plants and acute gamma rays of the seeds was effective and resulted in the following lodging resistant mutants:

1   The mutant RH 3112 induced by chronic gamma rays from the Kenyan cultivar with very erect and strong stems, thick shoots and secondary shoots arising from the base.
2   The mutant RH 3119 induced by acute gamma rays, also from the Kenyan cultivar, with dwarf like stem and shoots.

References

Agrios, N.G. (1969) *Plant Pathology*, Academic Press, New York and London.
Allard, R.W. (1960) *Principles of Plant Breeding*, J. Hilley & Sons Inc., London.
Anonymous (1961) *Yearbook of Agriculture*, US Dept. of Agriculture, Fisheries and Food, Washington, USA.

Anonymous (1968) *Review of Applied Mycology, Plant Host–Pathogen Index*, Commonwealth Mycological Institution, Vols. 1–40, p. 410, Kew, Surrey, England.

Anonymous (1970) *Short Term Leaflet 60*, Ministry of Agriculture, Fisheries and Food, Washington, USA.

Anonymous (1996) Common catalogue of varieties of agricultural plant species. *Official J. European Communities*, **39**, C 272 A, p.45.

Auerbach, C. (1961) Chemicals and their effects. *Proc. Symp. on Mutation and Plant Breeding*, Cornell, Nov.–Dec., 1960.

Awadala, M.Z., El-Gedaily, A.M., El-Shamy, A.E. and El-Aziz, K.A. (1980) Studies on some Egyptian food, Part I, Biochemical and biological evaluation. *Z. Ernährungswiss.*, 19(4), 244–7.

Brim, C.A. (1973) Quantitative genetics and breeding. In B.E. Caldwell (ed.), *Soybeans. Improvement, Production and Uses*, Agronomy Monograph 16, Am. Soc. Agron. Madison, Wi., pp. 155–86.

Brunt, A. (1972) *Official Report to Bath University*, Glasshouse Crops Research, Virology Dept. Institute at Littlehampton, England.

Busbice, T.H., Hill, R.R.Jr. and Carnahan, H.L. (1975) Genetics and breeding procedures. In C.H. Hanson (ed.), *Alfalfa Science and Technology*, Amer. Soc. Agron. Inc. Publ., Madison., Wi., USA, 283–319.

Caldwell, B.E., Weber, G.R. and Byth, D.F. (1966) Selection value of phenotypic attributes in soy-beans. *Crop Sci.*, 6, 249–51.

Collins, G.B. and Petolino, J.G. (1984) *Application of Genetic Engineering to Crop Improvement*, Martinus Nijhoff, Dr. W. Junk Publishers, USA.

Cornish, M.A., Hardman, R. and Sadler, R.M. (1983) Hybridisation for genetic improvement in the yield of diosgenin from fenugreek seed. *Planta Medica*, 48, 149–52.

Dachler, M. and Pelzmann, H. (1989) *Heil- und Gewürzpflanzen, Anbau-Ernte-Aufbereitung*, Österreichischer Agrarverlag, Wien.

Darlington, C.D. and Wylie, A.P. (1945) *Chromosome Atlas of Flowering Plants*, George Allen & Unwin Ltd., London.

Dawidar, A.M. and Fayez, M.B.E. (1972) Thin-layer chromatographic detection and estimation of steroid sapogenins in fenugreek. *Fresenius' Z. Anal. Chem.*, 259(4), 283–5.

De Candolle, A. (1964) *Origin of Cultivated Plants*, Hafner, New York.

Del' Gaudio, S. (1952) Ricerche sulla biologia della *Trigonella*. *Ann. Sper. Agr.*, 6, 507–16.

Del' Gaudio, S. (1953) Ricerche sui consumi idrici e indugini sull' autofertilita del fieno greco. *Ann. Sper. Agr.*, 7, 1273–87.

Dixit, B.S. and Srivastava, S.N. (1977) Detection of diosgenin in the seeds of *Trigonella foenum-graecum* Linn. by GLC method. *Indian J. Pharm.*, 39(3), 62.

Dubinin, N.P. (1961) *Problems of Radiation Genetics*, Oliver and Boyd, London.

Duke, A.J. (1986) *Handbook of Legumes of World Economic Importance*, Plemus Press, New York and London.

Edison, S. (1995) Spices – research support to productivity. In N. Ravi (ed.), *The Hindu Survey of Indian Agriculture*, Kasturi & Sons Ltd., National Press, Madras, pp. 101–5.

Elujoba, A.A. and Hardman, R. (1985a) Fermentation of powdered fenugreek seeds for increased sapogenin yields. *Fitoterapia*, 56(6), 368–70.

Elujoba, A.A. and Hardman, R. (1985b) Incubation conditions for fenugreek whole seed. *Planta Med.*, 51(2), 113–15.

Evans, L.T. (1984) Physiological aspects of varietal improvement. *Proc. 16th Stadler Gent. Symp.*, Columbia, 121–46.

Fazli, F.R.Y. and Hardman, R. (1968) The spice fenugreek (*Trigonella foenum-graecum* L.). Its commercial varieties of seed as a source of diosgenin. *Trop. Sci.*, 10, 66–78.

Furry, A. (1950) Les cahiers de la recherche agronomique, 3, 25–317.

Gaul, H. (1961) Mutation and plant breeding. *Proc. Sympos. on Mutation and Plant Breeding*, Cornell, Nov.–Dec., 1960.

Gopal, S.K. and Maggon, T.A. (1971) Contribution to the physiology of *Trigonella* infected with *Peronospora trifoliorum*. *Biol. Plant.*, 13(5–6), 396–401.

Green, J.M., Sharma, D., Reddy, L.J., Saxena, K.B., Gupta, S.C., Jain, K.C., Reddy, B.V.S. and Rao, M.R. (1981) Methodology and progress in the I.C.R.I.S.A.T., Pigeonpea Breeding Program. *Proc. Intern. Workshop on Pigeonpeas*, Patancheru, Dec., 1980.

Haque, S.E. and Ghaffar, A. (1992) Efficacy of *Tichoderma* spp. and *Rhizobium meliloti* in the control of root-rot of fenugreek. *Pakistan J. Botany*, 24(2), 217–21.

Hardman, R. (1969) Pharmaceutical products from plant steroids. *Trop. Sci.*, 11, 196–222.

Hardman, R. (1978) Fenugreek – a multi-purpose legume. *Association of Applied Biologists*, Norwich, April, 1978.

Hardman, R. and Brain, K.R. (1970). The biochemical manipulation of the yield of steroidal sapogenin from harvested plant material. *Internationale f'r Arzneipflanzenforschung*, Vienna, July, 1970.

Hardman, R. and Brain, K.R. (1972) Variation in the yield of total and individual 25a- and 25b-sapogenins on storage of whole seed of *Trigonella foenum-graecum* L. *Planta Medica*, 21, 426–30.

Hardman, R. and Fazli, F.R.Y. (1969) The variation in sapogenin content of *Trigonella foenum-graecum* L. (fenugreek) with morphological part and stage of development. *20th International Congress of Pharmaceutical Sciences*, Federation Internationale Pharmaceutique, London.

Hardman, R. and Fazli, F.R.Y. (1972a) Methods of screening the genus *Trigonella* for steroidal sapogenins. *Planta Medica*, 21, 131–8.

Hardman, R. and Fazli, F.R.Y. (1972b) Labelled steroidal sapogenin and hydrocarbons from *Trigonella foenum-graecum* L. by acetate, mevalonate and cholesterol feeds to seeds. *Planta Medica*, 21, 188–95.

Hardman, R. and Jefferies, T.M. (1972) A combined column chromatographic and infrared spectrometric determination of diosgenin and yamogenin in fenugreek seed. *The Analyst*, 97, 437–41.

Hardman, R. and Stevens, R.G. (1978) The influence of N.A.A. and 2,4 D on the steroidal fractions of *Trigonella foenum-graecum* static cultures. *Planta Medica*, 34, 414–19.

Hidvegi, M., El-Kady, A., Lòsztity, R., Bákás, F. and Simon-Sarkadi, L. (1984) Contribution to the nutritional characterization of fenugreek (*Trigonella foenum-graecum* L.). *Acta Alimentaria*, 13(4), 315–24.

Jain, S.C. and Agrawal, M. (1987) Effect of chemical mutagens on steroidal sapogenin in *Trigonella* species. *Phytochemistry*, 26(8), 2203–6.

Jain, S.C. and Agrawal, M. (1994) Effect of mutagens on steroidal sapogenin in *Trigonella foenum-graecum* tissue cultures. *Fitoterapia*, 65(4), 367–75.

Jefferies, T.M. and Hardman, R. (1972) The infra-red spectrometric estimation of diosgenin and yamo-genin individually and as their mixtures. *Planta Medica*, 22, 78–87.

Joshi, S. and Raghuvanshi, S.S. (1968) B-chromosomes, pollen germination in situ and connected grains in *Trigonella foenum-graecum*. *Beitr. Biol. Pf.I.*, 44(2), 161–6.

Kamal, R., Yadav, R. and Sharma, G.L. (1987) Diosgenin content in fenugreek collected from different geographical regions of South India. *Indian J. Agric. Sci.*, 57(9), 674–6.

Khanna, P. and Jain, S.C. (1973) Diosgenin, gitogenin and tigogenin from *Trigonella foenum-graecum* tissue culture. *Lloydia*, 30(1), 96–8.

Khanna, P., Jain, S.C. and Bansal, R. (1975) Effect of cholesterol on growth and production of diosgenin, gitogenin, tigogenin and sterols in suspension cultures. *Indian J. Exp. Biol.*, 13(2), 211–13.

Kozlowski, J., Nowak, A. and Krajewska, A. (1982) Effects of fertilizer rates and ratios on the mucilage value and diosgenin yield of fenugreek. *Herba Polonica*, 28(3–4), 159–70.

Laxmi, V. and Datta, S.K. (1987) Chemical and physical mutagenesis in fenugreek. *Biol. Mem.*, 13(1), 64–8.

Laxmi, V., Gupta, M.N., Dixit, B.S. and Srivastava, S.N. (1980) Effects of chemical and physical mutagens on fenugreek oil. *Indian Drugs*, 18(2), 62–5.

Manha, S.K., Raisinghani, G. and Jain, S.C. (1994) Diosgenin production in induced mutants of *Trigonella corniculata*. *Fitoterapia*, 65(6), 515–16.

Marques de Armeida, J. (1940) Study of improvement of *fenugreek* (*Trigonella foenum-graecum*). *Agronomia Lusitana*, 2, 307–35.

Mohamed, E.S.S. (1983) *Herbicides in fenugreek* (*Trigonella foenum-graecum* L.) *with particular reference to dios-genin and protein yields*, PhD Thesis, Bath University, England.

Olsson, G. (1960) Some relations between number of seeds per pod, seed size and oil content and the effects of selection for these characters in *Brassica* and *Sinapis*. *Hereditas*, 46, 29–70.

Palti, J. (1959) *Oidiopsis* diseases of vegetable and legume crops in Israel. *Plant Diseases Report*, 43(2), 221–6.

Paroda, R.S. and Karwasra, R.R. (1975) Prediction through genotype environment interactions in fenugreek. *Forage Res.*, 1(1), 31–9.

Pasich, B. Terminska, K. and Beblot, D. (1983) Diosgenin and yamogenin in domestic Semen Foenugraeci. *Herba Pol.*, 29 (3–4), 203–9.

Petropoulos, G.A. (1973) *Agronomic, genetic and chemical studies of* Trigonella foenum-graecum L., PhD Thesis, Bath University, England.

Petropoulos, G.A. (1988) The twin pods near the top of stem of fenugreek (*Trigonella foenum-graecum* L.) as index selection for higher diosgenin content of seed. *Proc. 2nd Scient. Conf. in Genet. Improv. of Plants*, Saloniki, October, 1988 (in greek).

Petropoulos, G.A. (1990) The width of pod, the fluorescent, the size and the shape of seed, as index of selection associated with crude protein content of fenugreek seed (*Trigonella foenum-graecum* L). *Proc.3rd Greek Scient. Soc. Genet. Improv. of Plant Conf.*, Athens, Oct. 1990 (in Greek).

Poehlman, J.M. (1979) *Breeding Field Crops*, Avi Publ. Co. Inc., Westport, CT., 486 pp.

Prasad, C.K.P.S. and Hiremath, P.C. (1985) Varietal screening and chemical control foot-rot and damping-off caused by *Rhizoctonia solani. Pesticides*, 19(5), 34–6.

Raian, F.S., Vedamuthu, P.G.B., Khader, M.P.A. and Jeyarajan, R. (1991) Management of root disease of fenugreek. *South Indian Horticulture*, 39(4), 221–3.

Reid, J.S.G. and Meier, H. (1970) Chemotaxonomic aspects of the reserve galactomannan in leguminous seeds. *Z. Pflanzenphysiol.*, 62, 89–92.

Ries, S.K. and Everson, E.H. (1973) Protein content and seed size. Relationships with seedling vigour of wheat cultivars. *Agron. J.*, 65, 884–6.

Rouk, H.F. and Mangesha, H. (1963) *Fenugreek (Trigonella foenum-graecum L.). Its relationship, geography and economic importance*, Exper. Stat. Bull. No. 20, Imper. Ethiopian College of Agric. and Mech. Arts.

Roy, R.P. and Singh, A. (1968) Cytomorphological studies of the colchicine-induced tetraploid *Trigonella foenum-graecum. Genet. Iber.*, 20(1–2), 37–54.

Saleh, N.A. (1996) *Breeding and cultural practices for fenugreek in Egypt*, National Research Centre, Cairo (personal communication).

Serpukhova, V.I. (1934) Trudy, *Prikl. Bot. Genet. i selekcii Sen.* 7(1), 69–106 (in Russian).

Schroeder, W.T. and Provvidenti, R. (1971) A common gene for resistance to Bear Yellow mosaic virus and watermelon virus 2 in *pisum satirum. Phytoph.*, 61, 846–8.

Sharma, G.L. and Kamal, R. (1982) Diosgenin content from seeds of *Trigonella foenum-graecum* L. collected from various geographical regions. *Indian J. Botany*, 5(1), 58–59.

Singh, D. and Singh, A. (1974) A green trailing mutant of *Trigonella foenum-graecum* L. (Methi). *Crop Improvement*, 1(1–2), 98–100.

Singh, D. and Singh, A. (1976) Double trisomics in *Trigonella foenum-graecum* L. *Crop Improvement*, 3(1–2), 125–7.

Singh, R.R. and Raghuvanshi, S.S. (1980) Effect of D.E.S. in combination with D.M.S.O. on 2X and 4X *Trigonella foenum-graecum* L. *Indian J. Hortic.*, 37(3), 310–13.

Sinskaya, E. (1961) *Flora of cultivated plants of the U.S.S.R. XIII. Perennial leguminous plants, Part I. Medic, Sweet clover, Fenugreek*, Israel Programme for Scientific Translations, Jerusalem.

Stevens, R.G. (1974) *Trigonella foenum-graecum L. aseptic cell cultures and their steroids*, PhD. Thesis, University of Bath, England.

Stevens, R.G. and Hardman, R. (1974) Steroid studies with tissue cultures of *Trigonella foenum-graecum* L. using GLC. *Proc. 3rd Intern. Congress of Plant Tissue and Cell Culture*, Leicester, 1974.

Trisonthi, P., Baccou, J.C. and Sauvaire, Y. (1980) Trial to improve production of steroidal sapogenin by fenugreek (*Trigonella foenum-graecum* L.) tissue grown *in vitro. C.R. Seances Acad. Sci.*, Ser. D., 291(3), 357–60 (in French).

Tungland, L., Chapco, L.P., Wiersma, J.V. and Rasmusson, D.C. (1987) Effect of erect leaf angle on grain yield in Barley. *Crop Sci.*, 27, 39–40.

Ullah, M. (1982) Processing effects on protein quality of different legume seeds. *Pak. J. Agric. Res.*, 3(4), 252–8.

Vaitsis, Th. (1985) Creation of a new variety of fenugreek, named 'Ionia', resistant to *Sclerotinia sclerotiorum*, Fodder and Pastures Research Institute, Larissa, Greece (unpublished data).

Varshney, I.P., Vyas, P. and Beg, M.F.A. (1980) Fatty acid composition of five saponins containing seed oils. *J. Oil Technol. Assoc. India.*, 12(1), 20–1.

Walker, J.C. (1952) *Diseases of Vegetable Crops*, Mc Graw-Hill Book Co. Inc., London.

Woodworth, C.M. (1922) The extent of natural cross-pollination in soybeans. *J. Amer. Soc. Agron.*, 14, 278–83.

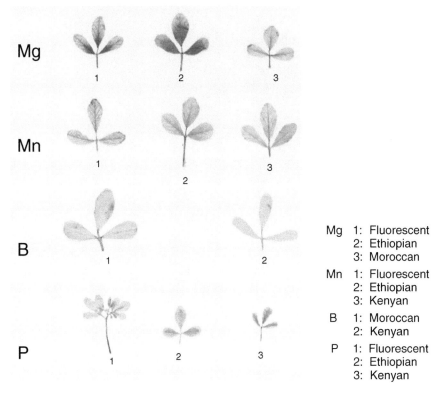

Mg 1: Fluorescent
    2: Ethiopian
    3: Moroccan

Mn 1: Fluorescent
    2: Ethiopian
    3: Kenyan

B   1: Moroccan
    2: Kenyan

P   1: Fluorescent
    2: Ethiopian
    3: Kenyan

Color Plate I (See Chapter 6, p. 108. Panagiotis Kouloumbis)

Figure 6.1 Leaves of different fenugreek cultivars with symptoms of mineral deficiencies
(Photo: G. Petropoulos).

Color Plate II (See Chapter 6, p. 111. Panagiotis Kouloumbis)

Figure 6.2 Boron deficiency symptoms in a hybrid fenugreek plant (Fluorescent × Kenyan)
(Photo: G. Petropoulos).

Color Plate III (See Chapter 6, p. 114. Panagiotis Kouloumbis)

Figure 6.3 Manganese deficiency symptoms on a fenugreek plant of the Ethiopian cultivar (Photo: G. Petropoulos).

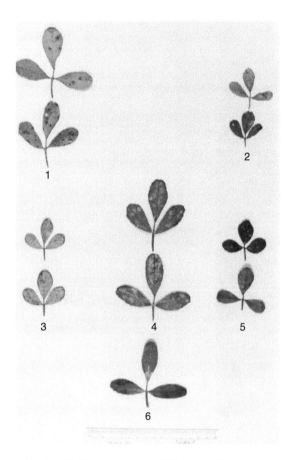

1: *Heterosporium* sp. in Fluorescent cultivar.

2: *Heterosporium* sp. in Ethiopian cultivar.

3: *Oidiopsis* sp. in Moroccan cultivar.

4: *Oidiopsis* sp. in Kenyan cultivar.

5: *Oidiopsis* sp. in Ethiopian cultivar.

6: Leaf miners in Kenyan cultivar.

Color Plate IV (See Chapter 7, p. 123. George Manicas)

Figure 7.1 Fenugreek leaves covered by different diseases (Photo: G. Petropoulos).

6 Nutrition and use of fertilizers

Panagiotis Kouloumbis

Introduction

In the old times, a fenugreek yield of 1 ton of seeds per hectare was considered very good, but nowadays yields of more than 2 tons per hectare are being obtained. The large yields of fenugreek are mainly dependent upon plentiful supplies of plant food in a form that fenugreek plants can readily use.

Meagre or sparse plant growth, slow growth and poor quality of grains and forage often indicate that there is a poor supply of plant food, which necessitates fertilization. The continuous cultivation of a soil by any rotation system results in a depletion of mineral nutrients. It is quite likely that one or more nutrients will become deficient even in fertile soils. Fenugreek is also sensitive to mineral deficiencies (Petropoulos, 1973).

Due to its sensitivity, especially in wet environmental conditions, it is very probable that the yellowing leaves of some fenugreek plants, described by Sinskaya (1961) as normal characteristics of some ecotypes, might be due to mineral deficiencies, particularly of boron (B), magnesium (Mg), manganese (Mn) and potassium (K) (Petropoulos, 1973).

Factors affecting nutrient needs

A number of factors must be taken into consideration when determining the amount of fertilizer that should be applied. As fenugreek is grown either as a fall or spring crop there are demands for nutrients under a wide range of environmental conditions. So, a good consideration of the effect of climatic and edaphic factors as well as cultural practices on the growth of fenugreek is necessary, in order to ensure adequate levels of all the essential elements throughout the growing period.

Soil

It is well known that three mechanisms, root interception, mass flow and diffusion govern the rate of supply of nutrients from the soil to the plant root (Oliver and Barber, 1966). In order to make the correct fertilizer recommendations a good knowledge of soil property is absolutely necessary. Soil capacity to retain nutrients and moisture varies widely. The response of fenugreek to phosphorus (P) and K is dependent upon the supplying content of the soil. Soil acidity also affects the availability of trace elements: iron (Fe), zinc (Zn), Mg and B, which are required for growth. The uptake of a nutrient depends on its concentration in the soil solution (Anonymous, 1990).

Level of yield

When fenugreek is managed for maximum seed yield it results in greater nutrient removal. As farmers obtain higher yields it will be necessary to increase the rates of maintenance application of fertilizers. Also it may become necessary to apply such elements that may not have been required in the past. So, increased yields are one of the major factors responsible for the increased use of fertilizers. Unfortunately, most experiments on fenugreek fertilization were conducted at yield levels that are low by present day standards. These results appear to be of questionable value and possibly misleading when making fertilizer recommendations for present day fenugreek growers.

Rainfall and temperature

Rainfall and temperature have a pronounced effect upon fertilizer response. Availability of some elements, like nitrogen (N) and P, is affected by temperature, since these nutrients become available from decomposing organic matter. Low soil temperature appears to limit the uptake of P and K (Smith, 1969). Phosphorous, K and calcium (Ca) concentrations were lower during growing seasons with high amounts of rainfall.

Intercropping system

Legumes, like fenugreek, are often richer competitors for Ca and Mg than grasses, while the opposite occurs in the case of K. Thus, when fenugreek is grown as a forage intercropped with barley or other grasses, in the case of K deficiency, the grain will tend to crowd out fenugreek unless more K fertilizer is applied from the beginning.

Stage of harvesting

As the concentration of many elements is higher in young plants (Anonymous, 1990), when fenugreek is grown as a forage early harvesting may result in the loss of more nutrients.

Nutrients removed annually

Petropoulos (1973) gives an analysis (Table 6.1) of fenugreek hay, as far as the percentage of removed amounts of main nutrients is concerned. According to this analysis, the approximate amounts of nutrients removed annually by the production of fenugreek hay per hectare (estimated hay yield: 2,000 kg/ha) is presented in Table 6.2. The approximate amounts of nutrients removed annually by fenugreek seed production/ha (estimated seed yield: 1,500 kg/ha) based on the analysis given by Duke (1986) is presented in Table 6.3, while Kouloumbis (1997) gives in Table 6.4 an analysis of plant nutrients removed by stalks and empty pods.

Soil acidity and liming

The need for lime can best be determined by a soil test. Lime is usually applied primarily to correct soil acidity. Although a pH value between 7.5 and 8.5 appears ideal for maximum fenugreek

Table 6.1 Proportion of main nutrients removed by fenugreek hay

No.	Nutrients	Percentage (f.m.b.)	ppm (f.m.b.)
1	N	2.60	
2	P	0.28	
3	K	1.72	
4	Ca	0.86	
5	Mg	0.14	
6	Mn		26
7	Cu		7.4
8	B		39

Table 6.2 Amount of nutrients removed annually by the production of fenugreek hay/ha (estimated yield of dry hay = 2,000 kg/ha)

No.	Nutrients	Removed amounts (kg)
1	N	52
2	P	5.6
3	K	34.4
4	Ca	17.2
5	Mg	2.8
6	Mn	0.052
7	Cu	0.0148
8	B	0.078

Table 6.3 Amount of nutrients removed annually by the edible portion of fenugreek seed production/ha (estimated yield of seed = 1,500 kg/ha)

No.	Nutrients	Removed amounts (kg)
1	N	67.7
2	P	5.4
3	K (Ash)	54
4	Ca	3.3
5	Fe	0.36

production the optimum pH for a fenugreek crop may vary considerably, depending upon soil characteristics such as texture, organic matter and lime in the subsoil.

Liming reduces the solubility of Fe, aluminum (Al) and Mg in the soil, while it can increase the availability of molybdenum (Mb) (Rhykerd and Overdahl, 1975). But overliming can decrease the availability of P and B.

Agricultural lime is a mixture of Ca or Ca and Mg and thus these nutrients are added in the soil when liming. Calcium promotes the root development of fenugreek and is essential for

Table 6.4 Analysis of plant nutrients in fenugreek stalks and empty pods

No.	Nutrients	Mature stalks	Mature pods without seeds
1	N%	1.00	0.675
2	P% (mg/100 g)	1.40	0.750
3	K% (g/100 g)	0.987	0.395
4	Ca (g/100 g)	0.51	0.63
5	Mg (g/100 g)	0.51	0.27
6	Cu (mg/kg)	7.15	3.10
7	Zn (mg/kg)	12.05	5.70
8	Mn (mg/kg)	15.75	7.85

nodulation and N fixation (Rhykerd and Overdahl, 1975), the fenugreek plant was found to be rich in Ca (Talwalkar and Patel, 1962).

The most important materials for liming are calcitic and dolomitic limestone. Dolomitic is often less effective than calcitic limestone. Lime is slow to react with soil and should be applied at least 1 year prior to sowing in strong acidic soils, and preferably not later than the fall of the year prior to sowing. Surface application without incorporation by plowing or disking is not recommended, due to the very slow movement of lime.

The recommended amount of lime is about 5 tons per acre. Half of it should be applied before plowing and half after plowing, followed by disking.

Nutrient macroelements

Nitrogen

Nitrogen is seldom applied to fenugreek crops that are pure and properly inoculated with *Rhizobium meliloti* (Del' Gaudio, 1962), except for a small amount called 'infantile nitrogen' about 20 kg/ha (Petropoulos, 1973; Heeger, 1989) at sowing time in soils that are low in organic matter. This is beneficial because it provides N for the first and rapid growth of fenugreek seedlings, until nodules form on the roots and the *Rhizobium* are able to fix large quantities of atmospheric N (Molgaard and Hardman, 1980). A liberal application of N fertilizer for fenugreek crop merely depresses the fixation of atmospheric N. In Egypt, it was found that in horse beans, when the soil contained 25–44 ppm mineral N, the N fixed amounted to 107 pounds per acre, where as when the initial content of mineral N was about 10 ppm, the N fixed rose to 114–154 pounds per acre (Rizk, 1966). In alfalfa, 18 pounds of N per acre was banded with P aided establishment, while 30 pounds proved detrimental (Rhykerd and Overdahl, 1975). In general, N fertilization in alfalfa tended to decrease the yield and stand and increase weeds (Rhykerd and Overdahl, 1975).

Recently, with the improved high yielding and high protein content of seed varieties of fenugreek and other legumes, the question has been raised as to whether nodule bacteria are capable of fixing adequate N for these cases. So, an increased interest has developed in studying the response of fenugreek to N fertilization in relation to these improved varieties.

The N content of healthy fenugreek plants is at least 2.5 percent (Table 6.1) and it is a basic constituent of the substances that are essential for protein synthesis. It is a constituent of chlorophylls and cytochrome enzymes, which are required for photosynthesis and respiration. Also many of the vitamins and alkaloids contain N. Nitrogen increases growth and defers maturity. It produces good leaves, aids stem development and gives a luxuriant dark-green color to plants, which is so desirable in growing crops.

The main source of N fertilization for fenugreek is limey nitrate ammonia for acid soils and sulfate ammonia for limey soils.

Phosphorous

Although the P content of a fenugreek plant is usually in the range of around 0.25 percent (Table 6.1), it participates in many vital life processes as the most important compounds containing P are nucleic acids and phospholipids, which play a vital role in photosynthesis, carbohydrate and protein synthesis and some coenzymes, necessary in oxidation–reduction reactions in all cells. It is quite mobile in the plant and moves from older to younger tissue when P is limited. Phosphorous hastens maturity of crops and hence lessens danger from frost damage in the fall, in wet and cold areas. It also aids in transferring substances from the stalk, leaves and other growing parts to the seed, making the grains plump and full. Phosphorous is absorbed very rapidly by young plants and in the case of alfalfa, when these tender plants have attained about 25 percent of their total dry weight, they may have accumulated, as much as 75 percent of their total phosphorus (Rhykerd and Overdahl, 1975).

Phosphates are relatively immobile in soil and the depth of its penetration appears to be related to the rate of P application and to soil texture. Alkaline and calcareous soils favour the solubility of P and in this respect liming can have a pronounced influence on availability of soil P.

As fenugreek is an annual crop, P fertilizer should be applied prior to sowing so it can be covered by plowing or disking. A depth of 10 cm in a normal soil was found to be an effective method of placement.

The most common source of P fertilizers are the ordinary and concentrated super phosphates, the latter containing between 40 and 50 percent of available P_2O_5.

The rate of application of P fertilizer depends mainly on the amount of available P in the soil and the yield level of fenugreek. Since the hay and seed production of fenugreek per hectare, according to Tables 6.2 and 6.3, would probably remove only about 5.6 and 5.4 kg/ha respectively, the rate of application of P appears quite small. However as the recovery of P fertilizer by a crop is generally low and usually ranging from 10–30 percent, the rate of application of P for high yields of fenugreek is often considerably greater than that which appears necessary, based on available P in the soil and crop removal.

Frequency of application does not appear to be critical with P in fenugreek, so the entire amount of fertilizers for each growing period is added once in the beginning, as it was mentioned above.

Symptoms of P deficiency are shown in Figure 6.1.

Potassium

The role of K affects a number of plant processes, like synthesis of carbohydrates, translocation of starch, synthesis of protein, control of activities of numerous essential mineral nutrients, neutralization of organic acids and activation of several important enzymes. Also K is essential for the formation of starch, sugar and cellulose, and when it is insufficient plants do not mature well.

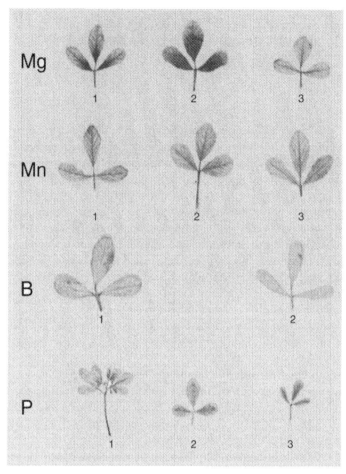

Mg 1: Fluorescent   B 1: Moroccan
    2: Ethiopian   2: Kenyan
    3: Moroccan
Mn 1: Fluorescent   P 1: Fluorescent
    2: Ethiopian   2: Ethiopian
    3: Kenyan   3: Kenyan

Figure 6.1 Leaves of different fenugreek cultivars with symptoms of mineral deficiencies (Photo:
      G. Petropoulos). (See Color Plate I.)

Potassium is present in fenugreek in a higher concentration than any other mineral element,
except N. The concentration of K in healthy fenugreek plants was found to be 1.72 percent
(Table 6.1), while in alfalfa early studies suggested that at the beginning a K concentration of
1–2 percent were adequate (Rhykerd and Overdahl, 1975). But more recent studies suggest that
a concentration of 2 percent or higher is necessary for maximum yield and longevity of the crop
(Rhykerd and Overdahl, 1975). So, K fertilizer is required in large amounts in many soils
poor in K for a successful fenugreek production, while in soils rich in K its addition was found
ineffective (Petropoulos, 1973).

The concept of critical percentage of mineral nutrients in plant tissue was developed by Macy, and is reported by Rhykerd and Overdahl (1975), but a number of factors, such as temperature and stage of development have a pronounced effect on K concentration in the plant.

Temperature affects K concentration in plants in alfalfa. The concentration of K under a cool temperature regime was 1.34 percent, as compared to 2.35 percent under warm temperature. These results suggest that, when the temperatures are cool, higher exchangeable K in the soil is required to ensure adequate K in the plant (Smith, 1969).

The stage of growth often influences concentration of K in plants to a greater extent than its availability in soils. The efficiency of K uptake appears to be closely related to the total root area of the plant (Oliver and Barber, 1966). A number of factors, such as soil, climate and yield level of fenugreek affect the rate and time of application of K fertilization. The determination of how much K the soil will supply and how much the fenugreek crop will remove make up the difference with the addition of K fertilization, since the hay and seed production of fenugreek per hectare, according to Tables 6.2 and 6.3, would probably remove about 34.4 and 72 kg/ha K respectively. This is a guide for estimating the amount of K fertilization of fenugreek.

On sandy loam soils with a pH value at least 6.5, 90 kg/ha of exchangeable K are recommended (Petropoulos, 1973), while Heeger (1989) suggests 80 kg/ha K_2O. Soil K is less in dry years, since the plant tries to feed in the subsoil, where the concentration of K is lower (Rhykerd and Overdahl, 1975). But the availability of K can also be reduced by excessive rainfall resulting in a lack of oxygen, which is necessary for respiration and K uptake (Rhykerd and Overdahl, 1975).

Losses of soil K occur due to leaching, erosion and cropping and these losses must be replaced frequently by the use of fertilizers. Muriate of potash (KCl) and potassium sulfate (K_2SO_4) are the two main sources of K fertilizer in fenugreek.

Potassium is a little more mobile in the soil than P, but much less mobile than nitrate. Potassium is absorbed on the base exchange complex, which accounts for its limited movement in soils.

Potassium fertilizer must be applied, either before, or at the time of sowing, followed by plowing or disking. Attention is to be given in case of the use of KCl, because of possible chloride injury to the seedlings (Rhykerd and Overdahl, 1975).

Dry matter, yield and total crude protein production increase, usually with an increasing rate of K fertilization. The influence of various types of fertilizers on the composition of the fenugreek seed was investigated. The use of K best increases the yield and the nutrient qualities with special effects on the oil content (Salgues, 1939). The feeding value of alfalfa increased as a result of K fertilization, mainly because of the increase in digestibility (Rhykerd and Overdahl, 1975).

Trace elements

Boron

Although the concentration of B in fenugreek is very low, B shortages can cause a serious reduction in crop yield.

The role of B in plants is very important as it is involved in many processes: pollen germination, cell division, water and carbohydrate metabolism and other processes. Carbohydrate translocation may be the most important function of B, since rapidly growing areas of the plant first exhibit deficiency symptoms.

Response from B is related more to yield than to quality. Soil organic matter and subsequent B release during decomposition is the basic source of this element as a nutrient (Rhykerd and Overdahl, 1975). Any difficulty in the bacterial action of decaying soil organic matter usually reduces the B supply.

In dry soil conditions, that push the plant to absorb nutrients from subsoil, which as known to be low in available B, its deficiency appears. Also a low soil pH inhibits bacterial activity and reduces B release from organic matter. Overliming reduces B availability, too. Any farming practice, like irrigation, that depletes soil organic matter, can magnify B deficiency. Leaching losses of B can be considerable, depending on soil texture and rainfall. Boron is not very mobile in the plant. In the case of shortage, B will be retained in the stem passing eventually to the leaves, and only if it is available will pass to the flowers and fruits (Tanaka, 1967; Sauchelli, 1969).

Furthermore, inoculation of fenugreek seeds with *Rhizobium* means the *Rhizobium*'s B requirements must also be satisfied (Petropoulos, 1973), because, as Hallsworth (1958) reports, Rhizobia may also have an absolute requirement for B, but lower than that of the host plant.

Although the actual amount of B needed by fenugreek plants is very small ranging around 40 ppm (Table 6.1), symptoms of B deficiency could occur if the soil is very low in B (Petropoulos, 1973). The B range of soils is usually 2–100 ppm (Chapman, 1966). In alfalfa, levels of B below 20 ppm in the top 6 in. of the plant indicated a deficiency of B (Anonymous, 1951). According to Chapman (1966), B deficiency in the early stages in many crops cannot be easily identified, except by leaf and soil analysis, which is frequently used. But B soil analysis is not a reliable measure of B availability, because less than 5 percent is in the available form. The lowest content of B, in which plants in water culture are showing B deficiency symptoms (no formation of pods), was 13 mg/g dry matter, and in this respect fenugreek is very similar to alfalfa, a plant known to have high B requirements (Molgaard and Hardman, 1980). Fenugreek possesses a high tolerance of excess B, as the very high B content of 62 mg/g dry matter in water culture fenugreek plants indicates this tolerance (Molgaard and Hardman, 1980).

The main symptoms of B deficiency, regardless of cultivar or variety, are failure of flowering or fertilization, decreased apical growth, small crisped leaves in a terminal rosette and a gradual yellowing of the lower leaves. In combination with low N, the B deficiency plants had yellow

Table 6.5 Boron deficiency symptoms for four fenugreek cultivars

Cultivars	Deficiency symptoms
Fluorescent	Leaves near the growing point are yellowed, lower leaves remain healthy, green color at the beginning but affected later. Lateral terminals are sometimes affected. Plants are stunted by a shortening of the terminal internodes resulting in rosetting. Flowers fail to form and buds appear as white or light brown dead tissue
Ethiopian	Young leaves turning yellow. Edges of some of these leaves later become bright red and then turn brown and die. Sometimes affects only the margins or the tip halves of leaves. The abnormal color spreads over the entire leaf surface including the veins
Kenyan	Leaves near a growing point are yellowed. Sometimes reddened. Lower leaves at first a healthy green, but later the symptoms are distributed over the entire plant. Plants are stunded by a shortening of the terminal internodes resulting in rosetting, which is characteristic. Affected leaves turning first dark brown and later light brown
Moroccan	Leaves of the younger portions of the plant are yellowed but later distributed over the entire plant. Plants are stunted by a shortening of the terminal internodes. Affected leaves die

Figure 6.2 Boron deficiency symptoms in a hybrid fenugreek plant (Fluorescent × Kenyan) (Photo: G. Petropoulos). (See Color Plate II.)

succulent leaves at a very early stage. High Ca and high N increased the demand for B (Molgaard and Hardman, 1980). The B deficiency symptoms for four breeding cultivars, as described by Petropoulos (1973), are tabulated in Table 6.5, while plants showing B deficiency symptoms are presented in Figures 6.1 and 6.2.

Materials that can be used to correct B deficiency are borax, which contains about 11 percent of B and boric acid. The correction is performed by foliar sprays using a solution containing about 2 percent $Na_2B_4O_7 \cdot 10H_2O$ with a suitable wetting agent, at a rate of approximately two-thirds of a fluid ounce per square yard. For soil application boric acid is usually used at a rate of 15 kg/acre (or 1–2 pounds of B per acre), which is adequate to limit B deficiency in fenugreek. Borax should not be used with ammonium salts because of a possible chemical reaction, whereas boric acid is compatible with it (Chapman, 1966). Soil applications give a longer correction than foliar sprays.

Magnesium

Magnesium is essential for photosynthesis (as a constituent of the chlorophyll molecule), carbohydrate metabolism and synthesis of oil. It is readily translocated from older to young tissue, in the case of Mg deficiency. There were strong indications of a lower content of fixed oils in seeds from plants showing symptoms of Mg deficiency. This may be due to the fact that Mg generally plays a role in oil formation (Anonymous, 1951). Also it was found that a higher content of Mg in fenugreek contributed to a higher fixed oils content in the seed (Petropoulos, 1973).

Soils that are developed on granites, sandstones and coastal sands are generally low in Mg, while those developed on dolomitic limestone and basic rock contain large amounts of Mg (Rhykerd and Overdahl, 1975). So, the content of Mg in soils varies widely. Magnesium deficiencies have developed due to many factors such as soil K (Rhykerd and Overdahl, 1975), the high content of alkaline soils in natrium (Rhykerd and Overdahl, 1975) and the continuous use of high calcitic limestone in soils low in Mg.

Dolomitic limestone is the main source of Mg. Potassium magnesium sulfate and $MgSO_4$ are mainly used to supply Mg, while magnesium chelate is used as a foliar spray, but it is expensive. The sulfate form of Mg is more soluble than dolomitic limestone.

Magnesium deficiencies mostly occur in soils containing less than 100 pounds of exchangeable Mg per acre (Rhykerd and Overdahl, 1975). Legumes normally contain two to three times as much Mg as do grasses (Chapman, 1966). It is recommended that some of the sulfate should be applied along with the initial liming to ensure adequate Mg. Chapman (1966) reports that the concentration of Mg in the mature leaves of plants without symptoms of deficiencies is 0.20–0.25 percent (m.f.b.), while for fenugreek plants it is 0.14 percent (Table 6.1), although according to Kansal and Pahwa (1979) fenugreek plants were found to be rich in Mg. Typical Mg deficiency symptoms, as described by Petropoulos (1973) for four breeding cultivars, are tabulated in Table 6.6, while fenugreek plants with Mg deficiency symptoms are presented in Figure 6.1.

It has been reported (Petropoulos, 1973) that fenugreek showed symptoms of Mg deficiency when other plants did not show such symptoms, and when plants such as fat hay grew in the margins of the experimental plots, according to Chapman (1966), it is an indicator of Mg deficiency. As Chapman (1966) states, the most common means of diagnosing Mg deficiency is by the use of visual symptoms. The appearance of a few leaves with characteristic Mg patterns is probably not serious enough to warrant the expense of corrective measures. The sufficient level for fenugreek hay, as has been reported previously, was found to be 0.14 percent, while for alfalfa the corresponding level also for dry hay is less than about 0.3 percent and for the top 6 inches of plants sampled prior to blooming it is 0.31–1.00 percent (Rhykerd and Overdahl, 1975).

Table 6.6 Magnesium (Mg) deficiency symptoms in four fenugreek breeding cultivars

Cultivars	Deficiency symptoms
Fluorecent	In early stages the area between the main veins of the leaves become pale green, later they turn a deep yellow except at the base of the leaf. Lower leaves are likely to be affected first. A later stage gives the general appearance of early maturity. A gradual yellowing from the margin and a bronzing over the entire leaf surface. Collapse of plants
Ethiopian	Central internal chlorosis and reddish brown marginal band. Collapse of the plants rarely occurs
Kenyan	Yellowing of broad margin of the leaf. The base and centre of the leaves and to some extent the veins remain green. In severe cases there is an almost complete yellowing of all leaves with a marked reduction in the growth
Moroccan	Central internal chlorosis of the leaves. Older leaves become chlorotic at the leaf margin and later in the midrib. Collapse of the plants rarely occurs

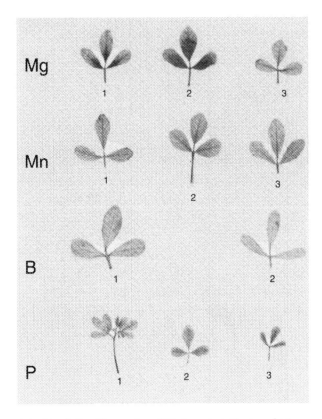

Mg 1: Fluorescent
2: Ethiopian
3: Moroccan

Mn 1: Fluorescent
2: Ethiopian
3: Kenyan

B 1: Moroccan
2: Kenyan

P 1: Fluorescent
2: Ethiopian
3: Kenyan

Color Plate I (See Chapter 6, p. 108. Panagiotis Kouloumbis)

Figure 6.1 Leaves of different fenugreek cultivars with symptoms of mineral deficiencies
(Photo: G. Petropoulos).

Color Plate II (See Chapter 6, p. 111. Panagiotis Kouloumbis)

Figure 6.2 Boron deficiency symptoms in a hybrid fenugreek plant (Fluorescent × Kenyan)
(Photo: G. Petropoulos).

Color Plate III (See Chapter 6, p. 114. Panagiotis Kouloumbis)

Figure 6.3 Manganese deficiency symptoms on a fenugreek plant of the Ethiopian cultivar (Photo: G. Petropoulos).

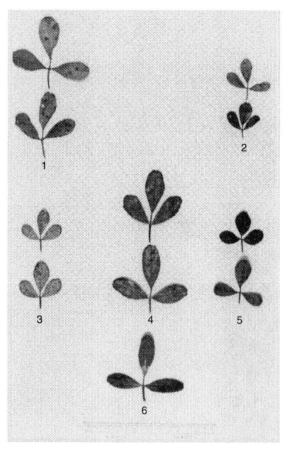

1: *Heterosporium* sp. in Fluorescent cultivar.

2: *Heterosporium* sp. in Ethiopian cultivar.

3: *Oidiopsis* sp. in Moroccan cultivar.

4: *Oidiopsis* sp. in Kenyan cultivar.

5: *Oidiopsis* sp. in Ethiopian cultivar.

6: Leaf miners in Kenyan cultivar.

Color Plate IV (See Chapter 7, p. 123. George Manicas)

Figure 7.1 Fenugreek leaves covered by different diseases (Photo: G. Petropoulos).

For correcting Mg deficiency, the application of a foliar spray with a solution containing 5–10 percent $MgSO_4 \cdot 7H_2O$ is recommended (this proportion depends on climatic conditions) with a wetting agent, at a rate of about two-thirds of a fluid ounce per square yard. For soil application, $MgSO_4$ is recommended to be broadcast at a rate of 40 pounds per acre, although for alfalfa in Ohio at a rate of 150–250 pounds per acre is recommended.

Manganese

Manganese along with Fe assists in chlorophyll synthesis and is involved in several oxidation–reduction systems. Excess of Mn can prevent the normal and reduced form of Fe in the plant.

Manganese deficiency in fenugreek, like alfalfa, can be produced by a neutral or alkaline pH, poor drainage or by biological factors (Graven *et al.*, 1965), as certain bacteria can oxidize the available Mn to the unavailable manganic form (Rhykerd and Overdahl, 1975).

Overliming a soil can produce Mn deficiency, and for this reason in soil low in Mn only moderate amounts of lime should be used. The low content of Mn in combination with neutral to alkaline soil favour the appearance of Mn deficiency in fenugreek (Petropoulos, 1973). In strong acidic soils Mn is reduced from the insoluble oxidized form to an exchangeable and available water-soluble form. Wallace (1951) stresses that Mn is more available in acid soils than in those that are neutral to alkaline in reaction.

The sufficiency range for fenugreek plants according to Table 6.1 is up to 26 ppm, while for alfalfa plants, sampled prior to bloom, it is from 26–100 ppm (Rhykerd and Overdahl, 1975). An excess of Mn causes a deficiency of Fe. Chapman (1966) reports that at least 3 ppm of exchangeable Mn in alkaline soils would have to be present for satisfactory crop production.

Very small differences were recorded from the Mn deficiency symptoms among the plants of four breeding cultivars of fenugreek and these are tabulated in Table 6.7, while fenugreek plants with Mn deficiency symptoms are presented in Figures 6.1 and 6.3.

Foliar sprays containing 4 percent $MnSO_4 \cdot 4H_2O$ with a suitable wetting agent are recommended for the correction of Mn deficiency symptoms, at a rate of about two-thirds of a fluid ounce per square yard, while for soil application manganese sulfate at a rate of about 20 pounds per acre (Petropoulos, 1973). Also about 50 pounds per acre of manganese sulfate (approximately 15–20 pounds Mn) in soils where the deficiency is known, is usually a satisfactory rate.

Table 6.7 Manganese (Mn) deficiency symptoms on four fenugreek breeding cultivars

Cultivars	Deficiency symptoms
Fluorescent	Symptoms are first seen in the young leaves. Light green to yellow leaves with distinctly green veins. Areas between the veins over the whole leaves become pale green and then pale yellow. In severe cases brown spots (necrotic areas) appear in leaves. Leaves drop off prematurely
Ethiopian	Symptoms as in fluorescent but less distinctly green veins
Kenyan	Symptoms as in fluorescent cultivar but the brown spots (necrotic areas) appear in higher proportion
Moroccan	Symptoms as in fluorescent

Figure 6.3 Manganese deficiency symptoms on a fenugreek plant of the Ethiopian cultivar (Photo: G. Petropoulos). (See Color Plate III.)

Zinc

Zinc plays an important role in several enzyme systems. Diminished growth and auxin concentration accelerates Zn concentration. Zinc deficient plants have a reduced water uptake. Although soils have an adequate Zn, in some of them there is a problem of availability, mainly in calcareous soils and in soils where high rates of P are applied.

The soluble forms of Zn are zinc sulfate and the chelated one, although the latter is very useful with a high Zn fixing capacity it is very expensive. The deficiency level is near 15 ppm for the whole alfalfa plant and a sufficiency range of 21–70 ppm for the top 6 inches of the plant sampled prior to blooming (Rhykerd and Overdahl, 1975). Five to 15 pounds per acre of Zn are generally applied as a soluble salt on soils where deficiencies are known.

Iron

Iron is involved in respiration since it is a constituent of the cytochromes. A deficiency of Fe is usually a consequence of low solubility rather than a mere absence. Iron is physiologically active in the ferrous state, but it is absorbed in the ferric state. The most common causes of Fe deficiency are overliming and the excess of Mn, which prevents the reduction of Fe in plant cells (Rhykerd and Overdahl, 1975).

Cold soil temperatures reduce the absorption of Fe. The sufficiency range is 30–250 ppm for alfalfa plants sampled prior to blooming (Rhykerd and Overdahl, 1975), while the fenugreek plant was found to be rich in Fe (Talwalkar and Patel, 1962). Tissue analysis may be the best indicator of Fe need. As Fe is poorly translocated a foliar application may correct deficient leaves, however new leaves may still be deficient.

Copper

Copper (Cu) is an enzyme activator and its role is complex and not clear. There are indications that Cu may be involved in the metabolism of root, protein and amino acids, in the rate of photosynthesis and in oxidation–reduction reactions.

Although the Cu content in soil varies with soil type (Rhykerd and Overdahl, 1975), most mineral and fine textured soils have enough native Cu content. Soil Cu is less available in alkaline than in acid soils. Some sandy and perhaps organic soils are poor in Cu.

Foliar rather than soil tests are usually better indicators for Cu need. The sufficiency range for fenugreek plants according to Table 6.1 is around 8 ppm, while for alfalfa in Ohio less than 11 ppm for plants sampled prior to blooming and showing Cu deficiency (Rhykerd and Overdahl, 1975). Copper deficient plants will respond to foliar feeding, but soil application is usually the most practical method of supplying Cu to fenugreek.

Copper sulfate, copper chloride and copper nitrate can be successfully used as fertilizers. On mineral soils 10 pounds of copper sulfate per acre of fenugreek are sufficient where deficiencies are known to occur, but on organic soils these amounts should be higher. But care must be taken to avoid toxic phenomena.

Combined fertilization

When considering commercial fertilizers, recognition should be given to the fact that the vegetative portion of fenugreek, like the other legumes, is high in K, P also is essential, but N should come from the atmospheric air.

The use of a nutrient extraction table is a good way to calculate the right NPK balance and the amount of fertilizer that should be applied, when no detailed information about nutrient requirements is available (Tables 6.1–6.3 for fenugreek).

For the case of fenugreek fertilization, some functional principles are reported below, based on general information for the cultivation of crops in a Mediterranean climate (Anonymous, 1990).

1   The uptake of a nutrient depends on its concentration in the soil solution and varies during the cropping cycle according to the amount and the type of mineral elements.
2   An excess of nutrients can have detrimental effects, such as phytotoxity or abnormal growth excesses. For example, excessive B results in plant death or excessive N can cause luxuriant leaf growth and a delay in maturity.
3   The application of nutrients to the soil in the exact proportions needed by fenugreek plants does not necessarily give good results, because they may not all be absorbed in the same way. For instance it is usual to apply more P than that extracted by the plants.
4   The application of nutrients should be proportional to plant uptake, to avoid any antagonism between nutrients. An example is the detrimental effect of high K application on Mg absorption, which is well known.
5   When saline water is used for irrigation, its nutrient content in certain conditions may be important with regard to plant nutrition. This is particularly true if the irrigation water has a high content of Ca, Mg, B or sulfur (S), as the example shown below.

Nutrient absorption by fenugreek plants is difficult to control because many soil factors are involved, for example: pH, temperature, exchange capacity, salinity and water supply. However, two methods may be used to build a fertilization program with sufficient precision: soil analysis and plant leaf analysis.

If *soil analysis* is carried out, using water as the extraction solvent instead of ammonium lactate, information about the content of nutrients in the soil solution is obtained but not about the potential nutrient reserve. It is possible, therefore, to make an accurate estimate of the amounts readily available to fenugreek plants. It is also possible to test the fertilization program by knowing the nutrient balance of the soil.

Leaf analysis is complementary to soil analysis for checking the nutrient composition of plants. However, the amounts of fertilizer that must be applied to obtain the correct leaf content vary widely, depending on growing conditions. So it is necessary to adjust a fertilization program to each fenugreek crop and region. This means that it is difficult for growers to use fertilization recommendations in relation to leaf analyses that have been established in other countries but not tested at home, or under similar climatic or growing conditions.

From the above, some general conclusions can be derived for fenugreek fertilization. They are the following:

- Water salinity must generally be controlled, particularly if soil drainage is incomplete.
- If the irrigation water has an intense alkaline character, serious P precipitation problems can be presented.

In order to reduce salt accumulation in the soil, small amounts of fertilizers are applied frequently rather than large quantities at longer intervals.

The great bulk of the fertilizers generally consist of N, phosphoric acid and K, either alone or in combination.

Complete commercial fertilizers of a 2-12-4 or 2-16-6 formula are effective. They should be used at a rate of 200 or 300 pounds per acre, applied either broadcast before sowing or as only a small amount of 50–75 pounds that can be drilled in the row with the fenugreek seed. Rathore and Manohar (1989) found that, in a winter crop of fenugreek on loamy sand, seed and straw yields were higher with 20 kg N/ha and 50 kg P/ha. Acid soils should be limed before they are seeded with fenugreek.

Band placement of fertilizers (N, P and K) is superior to a broadcast application and the same was found for other legumes (Rhykerd and Overdahl, 1975). N, P and K fertilizers improved the yield, while N and K improved the quality of hay used in fenugreek (Pareek and Gupta, 1981).

Nitrogen, P and K fertilizers had a beneficial effect on the fenugreek seed yield, while N and K improved the quality of fenugreek hay (Salgues, 1938b).

Pareek and Gupta (1981) reported that N and P application had a beneficial effect on fenugreek nodulation, while any direct relationship between N and P fertilizers and diosgenin content of fenugreek seed did not appear.

In pot experiments with fenugreek, it was found that the highest seed yield was obtained from the double N, P and K rates combined with Ca and Mg application (Golez and Kordana, 1979). Crops showed the highest requirement for N and K, lesser for Ca and least for P (Golez and Kordana, 1979).

Kozlowski *et al.* (1982) in pot experiments with fenugreek also found that seed yield was highest when the N–P–K rate was doubled. Mg addition increased the effect of the doubled N rate, but the highest seed yields were obtained when Ca was also added. The addition of Ca alone without Mg had a more positive effect on seed yield than addition of Mg alone. The average mucilage value was highest when Ca and Mg were added at doubled N rates. When Ca and Mg were added at doubled K rates the mucilage value decreased, while doubled K alone yielded the lowest diosgenin concentration. Without Ca and Mg the diosgenin concentration increased most when the N rate was doubled. A negative relation between N uptake and diosgenin content was observed.

As is generally known the health of a plant is expressed by the sum of its NPK contents in a given period. The lower mineralization of the green parts at the end of a vegetative cycle coincides with the optimum of change. Salgues (1938a) reported that healthy fenugreek plants at the end of a vegetative cycle had the lowest total of N, P and K, when either no fertilizer or K alone was used, followed in order by the use of complete N–P–K, N alone and P alone, while in other tested plants this order of fertilizers is different.

The use of high purity fertilizers does not supply enough to minor elements, so that the specific application of micronutrients becomes essential under these circumstances. The best way to achieve this is by foliar spraying, since incorporation in the soil can give very uncertain results for problems of precipitation and uptake.

Foliar feeding, in United Kingdom conditions, with a concentrated solution of trace elements containing 4% $MnSO_4 \cdot 4H_2O$ + 10% $MgSO_4 \cdot 7H_2O$ + 2% $Na_2B_4O_7$ with a wetting agent at a rate of two-thirds of a fluid ounce per square yard were used occasionally from when the first pods had formed until the beginning of September, with good results (Petropoulos, 1973). This early interruption of feeding in September took place to allow fenugreek to ripen.

Hardman (1980) suggests a fertilization of 20 units N, 50 units P and 50 units K (25, 63 and 65 kg/ha) to the seed bed and at 1–3 true leaf stage 70 units N (88 kg/ha element). While Hardman (1979) for feeding with trace elements, based on the trial growing of fenugreek in the United Kingdom, suggests for Mg 10 kg of element/ha, used as $MgSO_4 \cdot 7H_2O$, Mn 10 kg of element/ha used as $MnSO_4 \cdot 4H_2O$ and B 2.5 kg of element/ha used as H_3BO_3, making a solution of the first two salts in cold water, dissolving the boric acid in boiling water, mixing the solutions and spraying onto the land (avoid the use of a solution of borax, as this is incompatible with the solution of manganese sulfate).

The application of 140 kg N/ha as ammonium sulfate or ammonium nitrate, 50 kg P_2O_5/ha as calcium phosphate and 60 kg K_2O/ha as potassium sulfate, is a common practice in fenugreek fertilization in Egypt (Saleh, 1997). The recommended fertilization rate of fenugreek in Poland is 20–30 kg N/ha, 60–70 kg P_2O_5/ha and 80–100 kg K_2O/ha, 3–4 days before sowing (Anonymous, 1987). While in Germany and Hungary a similar opinion prevails: up to 20 kg N/ha (=100 kg/ha Calcium ammonium nitrate), 40–60 kg P_2O_5/ha (=270–400 kg/ha Thomasphosphate) and 80 kg K_2O/ha (=200 kg/ha K 40 percent) (Máthé, 1975; Heeger, 1989).

Hardman (1981), in notes issued for guidance and which cannot be taken as definitive, recommends in the case of fenugreek forage production for hay or silage, seed bed dressings with 18 and 50 units each of P and K, followed at the 3–4 true leaf stage by 70 units N and 15 units each of P and K, on soils that have an average nutritive situation.

References

Anonymous (1951) *Hunger signs in crops*, Amer. Soc. of Agronomy.

Anonymous (1987) *Kozieradka pospolita (Trigonella foenum-graecum L.) – Rodzina: Motylkowe (Papilionaceae)*, Instytut Roślin I Przetworów Zielarskich, W. Poznaniu, Zrzeszenie Przedsiebiorstw Przemyslu Zielarskiego 'Herbapol', 3 Str.

Anonymous (1990) *Protected cultivation in the Mediterranean climate*, F.A.O. Plant Production Protection, Paper No. 90, Rome, F.A.O. of the UN, pp. 313.

Chapman, H.D. (1966) *Diagnostic Criteria for Plants and Soils*, University of California, Riverside California.

Dachler, M. and Pelzman, H. (1989) *Heil- und Gewürzpflanzen, Anbau – Ernte – Aufbereitung*, AV – Berater, Österreichischer Agrarverlag, Wien.

Del' Gaudio, S. (1952) Il fieno greco, forragera del colle et del monte. *Ital. Agric.*, 89, 127–36.

Duke, A.J. (1986) *Handbook of Legumes of World Economic Importance*, Plenum Press, New York and London.

Golez, L. and Kordana, S. (1979) Effect of nitrogen, phosphorous and potassium doses, as well as magnesium and calcium fertilisation on a crop yield and uptake of mineral nutrients by *Trigonella foenum-graecum*. *Herba Pol.*, 25(2), 121–31.

Graven, E.H., Atoe, O.J. and Smith, D. (1965) Effects of liming and flooding on manganese toxicity in alfalfa. *Soil Sci. Soc. Amer. Proc.*, 29, 702–6.

Hallsworth, E. (1958) *Nutrition of the Legumes*, Butterworths Scient. Publ., London.

Hardman, R. (1979) Notes on the trial growing of fenugreek in the United Kingdom, Bath University, England (unpublished data).

Hardman, R. (1980) Fenugreek – a multi-purpose annual legume for Europe and other countries. *Cereal Unit Publication*, Royal Agricultural Show, Stoneleigh, UK.

Hardman, R. (1981) Fenugreek trials National Seed Development Organization Limited, Cambridge, England (unpublished data).

Heeger, E.F. (1989) *Handbuch des Arznei- und Gewürzpflanzenbaues*, Harri Deutsch Verlag, 2.Repr., Frankfurt/M.

Kansal, V.K. and Pahwa, A. (1979) Utilisation of magnesium from leafy vegetables and cereals. Effect of incorporation of skim milk powder in the diets. *J. Nutr. Diet.*, 16(12), 453–9.

Kozlowski, J., Nowak, A. and Krajewska, A. (1982) Zmiany wartosci śluzowej oraz zawartości i wydajności diosgeniny w nasionach kozieradki pospolitej (*Trigonella foenum-graecum* L.) pod wpływem zróżnikowanego nawożenia. *Herba Polonica*, 28(3–4), 159–70.

Kouloumbis, P. (1997) Analysis of fenugreek stalks and pods for plant nutrients, Athens Soil Science Institute (unpublished data).

Máthé, I. (1975) *A görögszéna (Trigonella foenum-graecum L.)*, *Magyarország Kult.*, III/2, Kultúrflóra 39., Akadémiai Kiadó, Budapest.

Miller, J.I. (1969) *The spice trade of the roman empire 29 B.C. to A.D. 641*, Clarendon Press, Oxford.

Molgaard, P. and Hardman, R. (1980) Boron requirements and deficiency symptoms of fenugreek (*Trigonella foenum-graecum*) as shown in a water culture experiment with inoculation of *Rhizobium*. *J. Agric. Sci. Camb.*, 94, 455–60.

Oliver, S. and Barber, S.A. (1966) An evaluation of the mechanisms governing the supply of Ca, Mg, K and Na to soybean roots (*Glycine max*). *Soil Sci. Soc. Amer. Proc.*, 30, 82–6.

Pareek, S.K. and Gupta, R. (1981) Effect of fertiliser application on seed yield and diosgenin content in fenugreek. *Indian J. Agric. Sci.*, 50(10), 746–9.

Petropoulos, G.A. (1973) *Agronomic, genetic and chemical studies of Trigonella foenum-graecum L.*, PhD. Thesis, Bath University, England.

Rathore, P.S. and Manohar, S.S. (1989) Effect of date of sowing, levels of nitrogen and phosphorous on growth and yield of fenugreek. *Madras Agric. J.*, 76(11), 647–8.

Rhykerd, C.L. and Overdahl, C.J. (1975) Nutrition and fertilizer use. In C.H. Hanson (ed.), *Alfalfa Science and Technology*, Amer. Soc. Agric. Inc. Publ., Ma., Wi., USA, 437–68.

Rizk, S.G. (1966) Atmospheric nitrogen fixation by legumes under Egyptian conditions. II. Grain legumes. *J. Microbiol. U.A.R.*, 1(1), 33–45.

Saleh, N.A. (1997) *Breeding and cultural practices for fenugreek in Egypt*. National Research Center, Cairo (personal communication).

Salgues, R. (1938a) Mineralization of the green parts (of plants) as a function of the application of fertilizers. *Bull. Assoc. Franc. Étude Sol*, 4, 36–44.

Salgues, R. (1938b) Studies of plant physiology. *Rev. Gen. Sci.*, 49, 238–42.

Salgues, R. (1939) Fenugreek, *Trigonella foenum-graecum* L. *Bull. Sci. Pharmacol.*, 64, 77–89.

Sauchelli, V. (1969) *Trace Elements in Agriculture*, Van Nostrand Reinhold, London, p. 248.

Sinskaya, E. (1961) *Flora of cultivated plants of the U.S.S.R. XIII. Perennial leguminous plants. Part I. Medic. Sweet clover. Fenugreek*, Israel Programme for Scientific Translations, Jerusalem.

Smith, D. (1969) Influence of temperature on the yield and chemical composition of 'vernal' alfalfa at first flower. *Agronomy Journal*, 61, 470–2.

Talwalkar, R.T. and Patel, S.M. (1962) Nutritive value of some leaf proteins. I. Amino-acid composition of *Trigonella foenum-graecum and Hibiscus cannabinus. Ann. Biochem. Exptl. Med.*, 22, 289–94.

Tanaka, H. (1967) Boron absorption by crop plants as affected by other nutrients of the medium. *Soil Science and Plant Nutrition*, 13(2), 41–4.

Wallace, I. (1951) *The Diagnosis of Mineral Deficiencies in Plants by Visual Symptoms*, Her Majesty's Stationary Office, London.

7 Pests and diseases

George Manicas

Although generally fenugreek is little subject to pest and fungal diseases (Sinskaya, 1961), a number of investigators have reported the appearance in fenugreek crops of some pest enemies and fungal, bacterial and viral diseases.

Pests

Fenugreek appears very resistant to attacks by insects and animal enemies and no serious damage in the plants has been recorded in the literature. It is also characteristic that in stored seeds of fenugreek, more than 10 years without any treatment, one did not notice any attack.

The peculiar smell that possesses the fenugreek plants and seeds may be a possible factor for their resistance to the attack of insects. The fact that dry fenugreek plants and seeds are mainly used as insect repellent to protect the grains from attacks of insects (Chopra *et al.*, 1965), may be connected and confirms partly the above hypothesis. The major pests that have been recorded as attacking fenugreek are presented in Table 7.1.

Table 7.1 The main pest enemies reported to attack fenugreek plants

Pest enemies	References
Adelphocoris lineolatus *Myzodes persicae* *Macrosiphon solanifolii* *Myzocallidium riehmi* *Agriotes ustulatus* *Asyrtosyphon pisum* *Agromyza frontella* *Agromyza nana* *Terias hecabe* *Plodia interpunctella*	Máthé, 1975
Chilo infuscatellus	Verum *et al.*, 1994
Tetranychus cucurbitae *Aphis craccivora* *Myzus persicae*	Duke, 1986
Rabbits Hares Game birds	Hardman, 1979
Leaf miners	Petropoulos, 1973; Hardman, 1979

As fenugreek is highly palatable to rabbits, hares and game birds, Hardman (1979) recommends that in case of severe attacks by these two pests to net the land against them, especially for experimental plots, by using netting of width about 120 cm, so that 30 cm is placed horizontally in the ground extending away from the growing area, at a depth of 20 cm such that 70 cm is standing vertically above the soil level.

Very occasionally leaf-miners (see Figure 7.1) and leaf-rollers damage were reported at Bath, which were easily controlled with malathion or other up-to-date insecticide (Petropoulos, 1973; Hardman, 1979).

Diseases

The main diseases that have been recorded to attack fenugreek are presented in Table 7.2.

From the diseases shown in Table 7.2, those that cause serious damage to fenugreek, are as follows.

Collar rot (Rhizoctonia solani *Kuhn*)

Fenugreek suffers extensively with foot-rot and damping-off of disease caused by *R. solani*, in some areas of India (Hiremath *et al.*, 1976). Studies were conducted to screen several varieties and cultivars of fenugreek and different fungicides for their efficacy in controlling this disease (Prasad and Hiremath, 1985). The varieties TG-18 and UM-20 showed some tolerance, while none of the varieties tested showed complete resistance.

A lot of fungicides have been tried by several investigators to control *R. solani* (Hiremath *et al.*, 1978; Prasad and Hiremath, 1985).

In vivo studies on control of fungus with different methods of fungicidal application showed that Carbedazim gave the best results, as seed as well as dry soil mix fungicide, while Captan was more effective as a soil drenching (Prasad and Hiremath, 1985).

Hague and Ghaffar (1992) found that *Rhizobium meliloti*, *Trichoderma hanatum*, *T. harzianum* and *T. pseudokoningii* used as seed dressing or as a soil drench completely controlled infection by *R. solani* in 30- and 60-day-old plants.

Powdery mildew (Oidiopsis *sp.*)

Palti (1959) has described this disease on fenugreek in Israel and considers it one of the most important diseases to afflict fenugreek. Rouk and Mangesha (1963) report that in Ethiopia fenugreek is usually attacked by *Oidiopsis* sp., which does considerable damage to the plants. Petropoulos (1973) reports that fenugreek plants of his experimental plots were infected by the fungus *Oidiopsis* that at first, caused slightly raised blister-like areas on the young leaves that soon became covered with a grayish white, powdery fungus growth, while the older leaves were covered with a white superficial powdery bloom of fungus growth (Figure 7.1). Chupp and Sherf (1960) observed that the pathogen of powdery mildew does not grow well when weather is wet, while Agrios (1969) notices that powdery mildew is a very common disease for arid and semi-arid environments.

Although *Oidiopsis* sp. is not a seed-borne disease, Petropoulos (1973) found that infection in fenugreek plants was higher from seeds untreated with Benlate, than from treated ones. It would appear that Benlate gives some systemic protection to the seedlings against *Oidiopsis* sp. The same worker found that sprays with Dinocap (Karathane) with 8–10 oz active ingredient/acre in 100 gals with a low volume sprayer gave satisfactory control. Among the four breeding cultivars

Table 7.2 The major diseases reported to attack plants of certain species of the genus *Trigonella*

Species of the genus Trigonella	Diseases/pathogens	References
T. foenum-graecum	Rhizoctonia solani	Raian et al., 1991; Haque and Ghaffar, 1992
	Ascochyta sp.	Minz and Solel, 1959; Anonymous, 1968
	Cercosporina sp.	Minz and Solel, 1959; Anonymous, 1968
	Cercospora traversiana	Bremer et al., 1952; Leppik, 1959; Leppik, 1960
	Peronospora trigonelae	Palti, 1956; Anonymous, 1968
	Leveillula taurica	Palti, 1956; Palti, 1959
	Pseudoperiza medicaginis	Glaeser, 1961
	Peronospora trifoliorum	Gopal and Maggon, 1971
	Peronospora trigonellae	Palti, 1956; Ciccarone, 1952
	Erysiphe martii	Nagy et al., 1972
	Uronyces trigonellae	Nagy et al., 1972; Palti, 1956; Ubrizsy, 1965
	Heterosporium sp.	Petropoulos, 1973
	Macrophomina phaseolina	Haque and Ghaffar, 1992
	Sclerotinia trifoliorum	Petri, 1934
	Fusarium oxysporum	Borg, 1936
	Xanthomonas alfalfa	Anonymous, 1968
	Bean Yellow Mosaic Virus	Anonymous, 1968; Petropoulos, 1973
	Potato virus A	Schmelzer, 1967; Anonymous, 1968
	Cow pea mosaic virus	Vidamo and Conti, 1965; Anonymous, 1968
	Potato virus Y	Schmelzer, 1967
	Tobacco etch. virus	
	Wisconsin pea streak virus	Anonymous, 1968
	Pea mosaic virus	
	Soybean mosaic virus	Quantz, 1968; Schmelzer and Wolf, 1971
	Watermelon mosaic virus	Quantz, 1968; Schmelzer and Wolf, 1971
	Alfalfa mosaic virus	Quantz, 1968; Schmelzer and Wolf, 1971
	Tomato black ring virus	Quantz, 1968; Schmelzer and Wolf, 1971
	Clover vein mosaic virus	Quantz, 1968; Schmelzer and Wolf, 1971
T. polycerata	Uromyces ciceris-arietini	Payak, 1962
	Uromyces antuyllitis f. Trigonella	Anonymous, 1968
T. caerulea	Broad bean mosaic virus	Anonymous, 1968
	Tobacco necrosis virus	
	Colletotrichum trifolii	Anonymous, 1968
	Pseudoperiza medicaginis	Glaeser, 1961
T. cretica	Colletotrichum trifolii	Anonymous, 1968
T. suavissima	Uromyces striatus	Anonymous, 1968

evaluated for their resistance to the fungus the Fluorescent was found to be the most tolerant, while the Moroccan one of the most susceptible (Petropoulos, 1973).

Leaf spot (Ascochyta *sp.*)

This disease causes irregular spots on fenugreek leaves up to 6 mm in diameter, turning brown to black, assuming definite margins and often a zonate appearance. Affected leaves may die and

Figure 7.1 Fenugreek leaves covered by different diseases, namely: 1. *Heterosporium* sp. in Fluorescent cultivar; 2. *Heterosporium* sp. in Ethiopian cultivar; 3. *Oidiopsis* sp. in Moroccan cultivar; 4. *Oidiopsis* sp. in Kenyan cultivar; 5. *Oidiopsis* sp. in Ethiopian cultivar; 6. Leaf miners in Kenyan cultivar (Photo: G. Petropoulos). (See Color Plate IV.)

fall (Figure 7.2). Pods may also be infected and the fungus can enter the seeds. Infected seeds are characterized by the presence of round dark brown lesions. Infection from diseased seeds results in a rot starting at the point of the seed attachment and advancing up the stem and down the tap root. The stem lesion may extend to a point above the soil line and young fenugreek plants are killed. Walker (1952) also confirms the seed-borne nature of this disease in pea, where it is carried in infected seeds and its overwinters in infected plants debris. Under UK conditions this fungus in field beans is favoured by cool moisture situations and rapid spread can occur during periods of rain, while in dry weather the disease may be confined to the lower part of the plant (Anonymous, 1970). Petropoulos (1973) reports that Benlate treatment of fenugreek seeds

Figure 7.2 Fenugreek plants affected by the fungus *Ascochyta* sp., where the leaves have died and fallen (Photo: G. Petropoulos).

protected the plants from primary infections, while frequent foliar sprays with Benlate protected them from secondary infections. But in weather conditions favourable to the spread of the disease, plants either from healthy seeds or from those treated with Benlate, may prove unable to control the *Ascochyta*, as low levels of disease can rapidly build up to produce an epidemic.

Variation in the sensitivity to attack by the *Ascochyta* sp. was found among four breeding cultivars, where Morrocan cultivar was proved the most susceptible, while the Fluorescent and Ethiopian ones the most tolerant (Petropoulos, 1973).

Pod spot *(Heterosporium sp.)*

This disease was investigated and described in fenugreek for the first time by Petropoulos (1973), from the related literature it was found that only one species of *Heterosporium* was recorded as infecting legumes, namely *Heterosporium medicaginis*, described as new by Karimov (1956).

The symptoms of *Heterosporium* sp. in fenugreek as described by the first of the above investigators are dark brown or black spots with a dark olive velvety cover on the pods, and are seen at the third stage of pod development. These spots, at the beginning, are elongated transversely to the axis of the pod and as they spread on the surface of the pod become more rounded (Figure 7.3). The same spots also occur on the base of the stem, while these are very rare on the leaves (Figure 7.1). The mycelium of the fungus is not buried deeply in the epidermis of the pod and the stem and appears to extend only into the first layers of the cells. There are no indications of it entering the seeds. Pirone *et al.* (1960) mention that the *Heterosporium* fungi are generally seed borne and hot-water treatment is a standard practice among seedsmen. Petropoulos (1973) believes that the contamination of fenugreek seed by this fungus takes place only during threshing. The same investigator reports that there are indications that the inoculation of the fenugreek

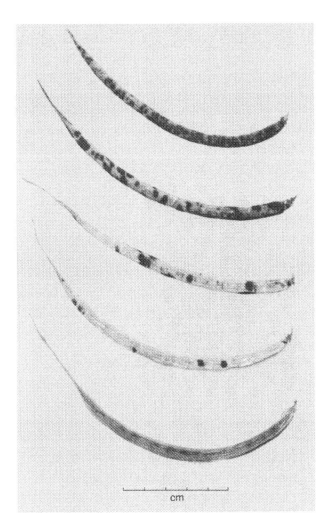

Figure 7.3 Dark brown and black spots of the fungus *Heterosporium* sp., spread on the surface of the fenugreek pods (different stages of disease development: upper: severely infected, lower: healthy) (Photo: G. Petropoulos).

seeds with *Rhizobium* increases the sensitivity of the plants to attacks by *Heterosporium* sp. and this may be due to the tenderness of the inoculated plants, as this fungus does not seem to have a high penetration ability.

Variation in the sensitivity to be attacked by this fungus was found among four evaluated breeding cultivars, the Ethiopian and Fluorescent cultivars were proved the most susceptible, while the Kenyan and the Moroccan were the most tolerant.

Bean Yellow Mosaic Virus (BYMV)

This virus is common in legumes, including fenugreek. According to Hill (1972) it is readily transmitted by many aphid species and is non-persistent, making control by aphicides difficult. On some legumes the virus has been recorded as being seed transmitted, although no actual record of this in fenugreek has been made. The prevention of virus infections in plants is difficult without isolating them from other virus hosts and from aphid vectors.

The main symptoms of BYMV infection in fenugreek is chlorosis and dwarfness (Petropoulos, 1973).

An experiment was carried out by the above worker, within the facilities of the Glasshouse Crops Research Institute at Littlehampton, England (Brunt, 1972), in order to investigate the severity of infection by BYMV and any tolerance to this virus by aphid and mechanical transmission, among four breeding fenugreek cultivars.

The conclusions drawn from these experiments are:

1   In the event of successful transmission of the BYMV on fenugreek plants severe symptoms of dwarfness and chlorosis will occur, so the disease is very serious.
2   The resistance of the fenugreek cultivars to transmission of the virus by aphids, which is the only mode of transmission in the field, is very favourable.
3   There are indications of some tolerance to this virus in the field in the case of Fluorescent and Ethiopian cultivars.

References

Agrios, N.G. (1969) *Plant Pathology*, Academic Press, New York and London.

Anonymous (1968) *Review of Applied Mycology. Plant Host–Pathogen Index*, Commonwealth Mycological Institution, Vols. 1–40, p. 410, Kew, Surrey, England.

Anonymous (1970) *Short term leaflet 60*, Ministry of Agriculture, Fisheries and Food, USA.

Borg, P. (1936) Report of the plant pathologist. *Rep. Insp. Agric. Malta*, 35, 53–61.

Bremer, H. *et al.* (1952) Beiträge zur Kentnisse der parasitischen Pilze der Türkei VII. *Rev. Fac. Sci. Univ. Istambul, Sér. B*, 227–88.

Brunt, A. (1972) *Data results*, Official Report to Bath University, Glasshouse Crops Research, Virology Dept. Institute at Littlehampton, England.

Chopra, R.N., Badhwar, R.L. and Ghosh, S. (1965) *Poisonous Plants of India*, Vol. 1. Indian Council of Agricutural Research, New Delhi.

Chupp, C. and Sherf, A.F. (1960) *Vegetable Diseases and Their Control*, Constable, London.

Ciccarone, A. (1952) Note fitopathologiche II. Segnalazione italiana della Trigonella (*Trigonella foenum-graecum* L.). *Ann. Sper. agr.*, N.S., 6, 165–8.

Duke, A.J. (1986) *Handbook of Legumes of World Economic Importance*, Plenum Press, New York and London.

Glaeser, G. (1961) Common leaf spot an autumn disease of lucerne. *Pflanzenarat*, 14(10), 88–9.

Gopal, S.K. and Maggon, T.A. (1971) Contribution to the physiology of *Trigonella* infected with *Peronospora trifoliorum*. *Biol. Plant*, 13(5–6), 396–401.

Haque, S.P. and Ghaffar, A. (1992) Efficiency of *Trichoderma* sp. and *Rhizobium meliloti* in the control of root rot of fenugreek. *Pakistan Journal of Botany*, 24(2), 217–21.

Hardman, R. (1979) Notes on the trial growing of fenugreek in the United Kingdom (unpublished data), Bath University, England.

Hill, S.A. (1972) *Official report to Bath University*, National Agricultural Station, Bristol, England.

Hiremath, P.C., Anilkumar, T.B. and Sulodmath, V.V. (1976) Occurrence of collar rot of fenugreek in Karnataka, India. *Curr. Sci.*, 45, 405.

Hiremath, P.C., Ponnappa, K.M., Janardhan, A. and Sundaresh, H.N., (1978) Chemical control of colar rot of fenugreek. *Pesticides*, 12, 30–1.

Karimov, M.A. (1956) Survey of fungal diseases of lucerne (*Medicago sativa*). *Not. Syst. Sect. Crypt. Inst. Sci.*, USSR, 11, 118–31.

Leppik, E.E. (1959) World distribution of *Cercospora traversiana*. *FAO Plant Prot. Bull.*, 8, 19–21.

Leppik, E.E. (1960) *Cercospora traversiana* and some other pathogens of fenugreek new to North America. *Plant Dis. Reptr.*, 44(1), 40–4.

Máthé, I. (1975) A *Görögszéna. Trigonella foenum-graecum* L. Magyarorszag III/2 Kulturfloraja 39, Akademiai Kiado, Budapest.

Minz, G. and Solel, Z. (1959) New records of field crop diseases in Israel. *Plant Dis. Rept.*, 43(9), 1051–9.

Nagy, F. *et al.*, (1972) *Cercospora traversiana* Sacc., a görögszéna (*Trigonella foenum-graecum* L.) új kórokozója Magyarországon és a védekezés Ichetöségei. *Herba Hung.*, 11(3), 53–60.

Palti, J. (1956) Parasites of fenugreek. *Hassedeh*, 37(3), 232–3.

Palti, J. (1959) Oidiopsis diseases of vegetable and legume crops in Israel. *Plant Dis. Report*, 43(2), 221–6.

Payak, M.M. (1962) Natural occurrence of Gram rust in uredial stage in *Trigonella polycerata* L., in Simla hills. *Curr. Sci.*, 31(10), 433–4.

Petri, L. (1934) Review of Phytopathological records noted in 1933, *Review of Applied Mycology*, Vol. 13, Kew, Surrey, UK.

Petropoulos, G.A. (1973) Agronomic Genetic and Chemical Studies of *Trigonella foenum-graecum* L., PhD Thesis, Bath University, England.

Pirone, P., Dodge, B. and Rickett, H. (1960) *Diseases and pests of ornamental plants*, 3rd edn, Constable and Company Ltd., London.

Prasad, C.K.P.S. and Hiremath, P.C. (1985) Varietal screening and chemical control foot-rot and damping-off caused by *Rhizoctonia solani. Pesticides*, 19(5), 34–6.

Quantz, L. (1968). *Leguminosen*. In M. Klinkowsky (ed.), *Pflanzliche Virologie*, II, 2, Akademie Verlag, Berlin.

Raian, F.S., Vedamuthu, P.G.B., Khader, M.P.A. and Jeyarajan, R. (1991) Management of root disease of fenugreek. *South Indian Horticulture*, 39(4), 221–3.

Rouk, H.F. and Mangesha, H. (1963) Fenugreek (*Trigonella foenum-graecum L.*). Its relationship, geogaphy and economic importance. *Exper. Station Bull.*, No. 20, Emp. Ethiopian Coop. of Agr. and Mech. Arts.

Schmelzer, K. (1967) Hosts of potato virus Y and Potato etch. virus outside of Solanaceae. *Phytopath. Z*, 60(4), 301–15.

Schmelzer, K. and Wolf, P. (1971) *Wirtspflanzen der Viren und Virosen Europas*, Barth Verlag, Leipzig.

Sinskaya, E. (1961) *Flora of cultivated plants of the U.S.S.R.. XIII Perennial Legumious plants: Part I. Medic. Sweet Clover, Fenugreek*, Israel program for Scientific Translations, Jerusalem.

Ubrizsy, G. (1965) *Növénykórtan II*. Akad. Kiadó, Budapest.

Verum, L.L. Suchita, S., Pandek, K.P. and Singh, S.B. (1994) Influence of companion cropping of spices on the incidence of early shoot borer (*Chilo infuscatellus*). *Indian Sugar*, 44(1), 21–2.

Vidamo, C. and Conti, M. (1965) Aphid transmission of a cowpea mosaic virus, isolated from cowpea in Italy. *Att. Acad. Sci.*, *Torino*, 99(6), 1041–50.

Walker, J.C. (1952) *Diseases of Vegetable Crops*, Mc Graw-Hill Book Co., Inc., London.

8 Weeds

C.N. Giannopolitis

Studies on weed interference and control in *Trigonella* spp. have been confined to one species, *T. foenum-graecum*, which is the main cultivated species of the genus around the world. Discussion in this chapter, therefore, will review results of research conducted so far in *T. foenum-graecum* grown mainly for seed production. Throughout the discussion, the name fenugreek, an internationally accepted common name of the species, is used as it is more convenient.

Weed interference

Although fenugreek, as a crop, grows and reaches maturity in a relatively short period (4–5 months), it is initially slow-growing and vulnerable to weed interference particularly during the seed germination and seedling establishment phases. It is therefore necessary that adequate control measures are applied to eliminate weed growth during these phases, if a good crop stand is to be obtained.

Weeds interfere with the growth of fenugreek seedlings mainly by competing with them for available nutrients and moisture and restricting available space. As well documented in many crops (Zimdhal, 1980), final yield reduction because of weed competition is mainly determined by:

1 The time and duration of competition. This means that yield reduction is greater the earlier the weeds germinate and the longer they are left to compete with the crop.
2 The relative crop/weed plant density. Yield reduction increases as the weed density (plants per square meter) becomes higher.
3 The relative (to the crop) competitive ability of the weeds present. Fast growing weeds that reach high fresh weight values in a short time are very competitive.

Weed competition in fenugreek, therefore, can be very strong if there is a heavy infestation by early-germinating annuals, or in the presence of highly competitive and fast-growing perennials. On the other hand, fenugreek takes good advantage if sown in a field cleared of perennial weeds and when placed in a crop rotation that reduces infestation from annuals.

From field trials in India (Tripathi and Govindra, 1993), it was concluded that the critical crop–weed competition period extends over the first 30 days after sowing of fenugreek. Weeds emerging during this period caused a yield reduction of 14.2 percent if they were removed soon and a reduction of 69 percent if they were left for the entire cropping season. Weeds emerging after the critical period of 30 days caused only a slight yield reduction (12 percent) and there was no significant advantage in increasing the weed-free period beyond the first 30 days.

Besides competition, certain weed species also have the potential to reduce fenugreek germination and growth through allelopathy, that is, by inhibitory chemicals that they release into the soil. Phenolic compounds and alkaloids that reduce seed germination or seedling growth of fenugreek have been detected in leachates from *Imperata cylindrica* L. Beauv. (Inderjit-Dakshini, 1991), *Argemone mexicana* L. (Leela, 1981) and other species. These results, however, do not allow any estimation of the final impact that allelopathy may have on yield under field conditions, and further research is needed.

Depending on the geographical region, the location, soil type and many other factors, a wide spectrum of weed species may be found in fenugreek crops. Both winter and spring species may be a problem. Of the winter species, plants of the *Cruciferae* family (e.g. *Sinapis* spp.) and other *Leguminosae* (e.g. *Melilotus* spp., *Trifolium* spp., etc.) can be very troublesome. Of the spring species, the early germinating broadleaves (e.g. *Chenopodium* spp.) and grasses (e.g. *Poa annua*, *Echinochloa crus-galli*, *Setaria* spp.) can be serious, especially in spring-sown fenugreek. Perennial species like *Convolvulus arvensis*, *Cyperus rotundus*, *Cynodon dactylon* etc., which are very difficult to control, create a very bad situation for the grower, if present.

Parasitic flowering plants may occasionally be a problem. *Orobanche indica* Ham. was found to parasitize the roots of fenugreek in India in fields of the Jaipur district where the weed density ranged from 8–32 plants per square meter (Bhargava *et al.*, 1976). *Orobanche crenata* Forsk., on the other hand, does not parasitize fenugreek neither is it induced to germinate by fenugreek root extracts (Khalaf, 1994).

Weed control

Prevention of weed competition during the critical period of the first 30–40 days after sowing should be the primary objective of any weed control program in fenugreek.

Field trials in India have revealed that two hand hoeings during the critical period are, under normal conditions, sufficient for a maximal seed yield. Maliwal and Gupta (1989) found that hand hoeing on the twentieth and fortieth days after sowing raised the seed yield to a level practically equal to that of the weed-free check. Compared to the unweeded check, the two-hoeing treatment more than doubled the yield and the increase was found to be connected with more pods per plant, more seeds per pod and a higher thousand-grain-weight of the seeds. Similarly, in field trials by Mandam and Maiti (1994), various weed control treatments increased the fenugreek seed yield from the unweeded check value of 0.88 t/ha to 0.96–1.2 t/ha. Hand weeding twice, 15 and 30 days after sowing, resulted in the highest seed yield.

Hand weeding is difficult and expensive and very seldom used in modern agriculture. However, by growing fenugreek as a row crop mechanical hoeing becomes a good alternative. A superficial soil disturbance, usually 2–3 times during the critical period, can effectively eliminate weeds between the crop rows, if performed at the right time with the proper equipment. Of course, it has to be supplemented with hand weeding on the rows. Other mechanical means (brushers, flamers, etc.) can also be used between the rows.

Herbicides

Herbicides are an effective means for weed control in most crops, with a better benefit–cost ratio than other methods. Some of the fundamental factors that must be considered when deciding on the use of a herbicide are selectivity to the crop, efficacy in controlling the weed species expected in the field and the risk of herbicide residues (above a permitted level) in the harvested product. Research in fenugreek, so far, is far behind a thorough examination of these factors and only few sound recommendations can be formulated on the basis of published data.

Richardson (1979) examined the tolerance of fenugreek to many herbicides applied pre- and post-emergence in pot experiments. Post-emergence herbicides that were well tolerated by fenugreek included bentazon, MCPB (Na salt), diclofop-methyl and alloxydim-Na. Pre-emergence herbicides with good selectivity to fenugreek included chlorthal-dimethyl, propyzamide, butam and propachlor as surface sprays and trifluralin, tri-allate and chlorpropham as soil-incorporated treatments.

Tolerance of fenugreek to trifluralin and other dinitroaniline herbicides has been further confirmed with field experiments, in which the efficacy evaluation of the herbicides was also made. Fluchloralin at 3.0 kg/ha was found to be the best treatment (following the hand-weeding treatment) by Mandam and Maiti (1994). Pendimethalin gave the best benefit–cost ratio in field trials by Maliwal and Gupta (1989). Tolerance to bentazon has also been confirmed with field experiments (Mandam and Maiti, 1994). Other herbicides may also be safe to be used for fenugreek, providing their selectivity is confirmed in the specific local conditions. Metamitron selectivity, for example, is marginal and seems to vary depending on the cultivar of fenugreek grown.

No selectivity problems are expected with the graminicides (e.g. diclofop, fluazifop-*P*, quizalofop-*P*, sethoxydim, clethodim, etc.), which can be very useful for the post-emergence control of annual and perennial grasses. A residue risk assessment is, however, needed with these herbicides, especially in cases where fenugreek is used as a fresh vegetable or as a forage plant, before their use is decided.

Another relevant aspect is the probability of damage to fenugreek by residues carried over in the soil from herbicides used in previous crops. In a study in Egypt, fenugreek was found to be the most susceptible crop out of six examined winter crops (wheat, barley, lentil, clover, broad bean) to atrazine residues in the soil. The high sensitivity of fenugreek is also expected with regard to residues from some of the sulfonylurea herbicides used in rotational crops.

Based on the best evidence available and the author's experience, the following practical recommendations can be made with regard to herbicide usage.

Pre-sowing treatments

The non-selective herbicides paraquat, glufosinate and glyphosate can be used before sowing to reduce weed density in the field. If the seed bed is prepared and preirrigated, well in advance, weeds will be forced to germinate before sowing and can be easily killed by spraying with the lowest recommended rates of the above herbicides. Glyphosate is also useful in reducing density of perennial weeds if used at higher rates during the period preceding that of fenugreek growing.

Pre-emergence treatments

A soil-acting herbicide that can selectively prevent emergence of weeds for at least a month would be suitable. The dinitroaniline herbicides trifluralin, fluchloralin and pendimethalin seem to be safe in most situations. The first two herbicides are applied shortly before sowing and are incorporated into the soil. Pendimethalin is usually applied to the soil surface soon after sowing but it can also be used as a pre-sowing incorporated treatment when dry conditions are expected. Other pre-emergence herbicides can also be used if they have been proven sufficiently selective to the crop under local conditions.

Post-emergence treatments

A great variety of weed species is usally found in fenugreek crops and none of the pre-emergence herbicides is sufficiently effective on all of them. The dinitroaniline herbicides, for example, are not effective on cruciferous weeds whereas other herbicides are weak on *Amaranthus* spp., *Chenopodium* spp. or grass species (Giannopolitis, 1981). A supplemental post-emergence treatment against escaping weeds may therefore be necessary.

Bentazon or MCPB, or a mixture of both, can be used against broadleaves and are usually effective if used properly. Other post-emergence herbicides can also be used if their selectivity has been established in the given local conditions.

A herbicide from the group of the specific graminicides (fluazifop-*P*, quizalofop-*P*, sethoxydim etc.) can be used against grass weeds provided that recommendations on the label for residue avoidance are followed. The mixing of these herbicides with other herbicides for simultaneous control of grasses and broadleaves, may reduce the efficacy of both and should be avoided (Giannopolitis, 1986).

Before using any of the herbicides mentioned in this chapter, local recommendations and restrictions should be considered carefully.

References

Abdel-Rahman, G.A. (1996) Susceptibility of certain winter crops to atrazine herbicide and detoxification by charcoal, organic manure and bioactive agents. *Ann. Agric. Sci. Moshtohor*, **34**, 733–41.

Bhargava, L.P., Handa, D.K. and Mathur, B.N. (1976) Occurrence of *Orobanche indica* on *Trigonella foenum-graecum* and *Physalis minima*. *Plant Dis. Rep.*, **60**, 871–2.

Giannopolitis, C.N. (1981) *Amaranthus* weed species in Greece: dormancy, germination and response to pre-emergence herbicides. *Annales Institut Phytopathologique Benaki*, **13**, 80–91.

Giannopolitis, C.N. (1986) Antagonistic interaction of herbicides on *Portulaca oleracea*. *Annales Institut Phytopathologique Benaki*, **15**, 77–80.

Inderjit-Dakshini, K.M.M. (1991) Investigations on some aspects of chemical ecology of cogongrass, *Imperata cylindrica* (L.) Beauv. *J. Chem. Ecol.*, **17**, 343–52.

Khalaf, K.A. (1994) Intercropping fenugreek with faba bean or Egyptian clover: prospects for *Orobanche crenata* control. In A.H. Pieterse, J.A.C. Verkleij and S.J. ter Borg (eds), *Biology and Management of Orobanche, Proceedings of the 3rd International Workshop on Orobanche and Related Striga Research*, Amsterdam, The Netherlands, Royal Tropical Institute, pp. 502–4.

Leela, D. (1981) Allelopathy in *Argemone mexicana* L. *Proceedings of the 8th Asian-Pacific Weed Science Society Conference*, pp. 401–4.

Maliwal, P.L. and Gupta, O.P. (1989) Study of the effect of four herbicides with and without applied phosphorous on weed control and seed yield of fenugreek (*Trigonella foenum-graecum* L.). *Trop. Pest Manage.*, **35**, 307–10.

Mandam, A.R. and Maiti, R.G. (1994) Efficacy of different herbicides for weed control in fenugreek (*Trigonella foenum-graecum* L.). *Environ. Ecol.*, **12**, 138–42.

Richardson, W.G. (1979) The tolerance of fenugreek (*Trigonella foenum-graecum* L.) to various herbicides. *Technical Report* No. 58, Agricultural Research Council, WRO, p. 31.

Tripathi, S.S. and Govindra, S. (1993) Crop-weed competition studies in fenugreek (*Trigonella foenum-graecum* L.). *Proceedings of the Indian Society Weed Science, International Symposium*, Hisar (India), 18–20 Nov. 1993, Vol. II, pp. 41–3.

Zimdahl, R.L. (1980) *Weed-Crop Competition – A Review*. International Plant Protection Center, Oregon State University, Corvallis, Oregon, p. 195.

9 Chemical constituents

Helen Skaltsa

Introduction

Trigonella foenum-graecum L., grown in many parts of Europe, Asia and Africa as a food (fresh green shoots, flour), spice (seeds, flour) and for use in native medicine, was already known by the ancient Egyptians and Greeks.

The greek name of the plant is "telis", which means green (Carnoy, 1959). The Romans learned from the Greeks that this plant of Oriental origin, used as a fodder, from which its name of "greek hay" is derived (André, 1956).

The biological and pharmacological actions of fenugreek are attributed to the variety of its constituents, namely: steroids, *N*-compounds, polyphenolic substances, volatile constituents, amino acids, etc.

Fenugreek seeds contain *c.* 6.2 percent moisture, 23.2 percent protein, 8 percent fat, 9.8 percent fiber, 26.3 percent mucilaginous material (see Chapter 3) and 4.3 percent ash. Whole grain is reported to contain (per 100 g of edible portion): 369 calories, 7.8 percent moisture, 28.2 g protein, 5.9 g fat, 54.5 g total carbohydrate, 8 g fiber, 3.6 g ash. Its flour contains 375 calories, *c.* 9.9 percent moisture, 25.5 g protein, 8.4 g fat, 53.1 g total carbohydrate, 7.1 g fiber, 3.1 g ash. Raw leaves contain 35 calories, *c.* 87.6 percent moisture, 4.6 g protein, 0.2 g fat, 6.2 g total carbohydrate, 1.4 g fiber, 1.4 g ash (Duke 1986).

Chemical constituents of other species, which have already been studied, are also described.

Steroids

Trigonella foenum-graecum L.

Common fenugreek is one of the few natural sources of the steroid sapogenin due to its seed content of diosgenin (Figure 9.1). The seeds have received extensive investigations by different research groups.

The C_{27} steroidal sapogenin diosgenin (Δ^5, 25α-spirostan-3β-ol) is of considerable economic importance to the pharmaceutical industry as a starting material for the partial synthesis of oral contraceptives, sex hormones and other medicinally useful steroids. Diosgenin has been extracted traditionally from the tubers of the Mexican and Asian species of yam, *Dioscorea*. However, an increased demand for raw steroid led the industries to look for an alternative source of diosgenin and other precursors.

Several investigators proposed fenugreek seeds as an alternative source for diosgenin (Marker *et al.*, 1947; Fazli and Hardman, 1968; Bhatnagar *et al.*, 1975). Hardman has proposed that fenugreek could be developed as a more widely grown multipurpose legume affording a cultivated source of diosgenin with its equally acceptable epimer, yamogenin (II) (Figure 9.1).

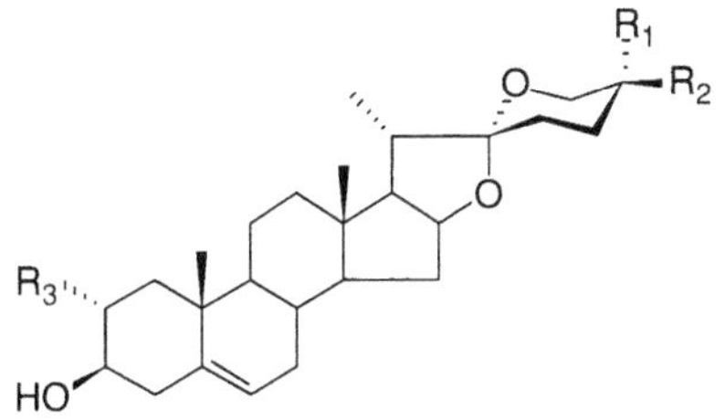

	R₁	R₂	R₃	
I	H	CH₃	H	Diosgenin
II	CH₃	H	H	Yamogenin
V	H	CH₃	OH	Yuccagenin
VI	CH₃	H	OH	Lilagenin

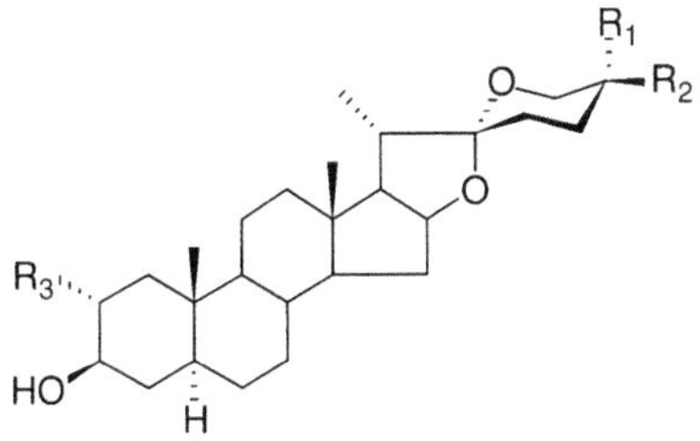

	R₁	R₂	R₃	
III	H	CH₃	H	Tigogenin
IV	CH₃	H	H	Neotigogenin
VII	H	CH₃	OH	Gitogenin
VIII	CH₃	H	OH	Neogitogenin

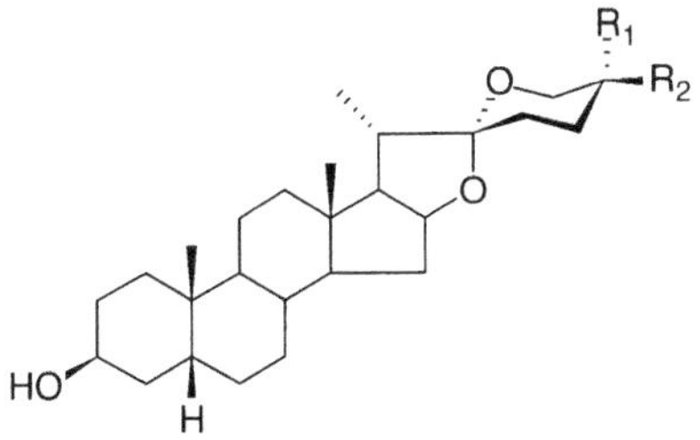

	R₁	R₂	
IX	H	CH₃	Sarsapogenin
X	CH₃	H	Smilagenin

Figure 9.1 Chemical structures of sapogenins.

The genins of fenugreek seed have been the subject of somewhat contradictory reports.

Soliman and Mustafa (1943) reported the presence of a steroidal sapogenin in the alcoholic extract hydrolysate of the fenugreek seed. Marker *et al.* (1943) in the course of plant studies for new sources of steroidal sapogenins extracted the same sapogenin from the seed and identified it as diosgenin (I). Shortly afterwards, Marker *et al.* (1947) described the sapogenin mixture, which they obtained from powdered fenugreek seed, as being made up mostly of diosgenin (yield about 1.0 g/kg dry seed) along with gitogenin (Figure 9.1) (5α, 25α-spirostan-2α, 3β-diol) (VII) (Figure 9.1) (0.1 g/kg dry seed) and traces of tigogenin (5α, 25α-spirostan-3β-ol) (III) (Figure 9.1). Soliman and Mustafa (1949) reported once again on the steroidal sapogenins of fenugrek seed, and confirmed Marker's findings with respect to the presence of diosgenin and gitogenin, but they did not mention tigogenin. Moreover, Soliman described another sapogenin he isolated in appreciable amounts from the mixture and assuming it to be new, named it trigonellagenin. Bedour *et al.* (1964), using defatted and powdered seed reported the isolation of diosgenin, gitogenin, tigogenin and a fourth product identical to 25α-spirosta-3,5-diene (*c.* 20 percent of the weight of diosgenin), which they suggested to be an artifact of diosgenin, produced during the acid hydrolytic processing of natural saponins, but they failed to find trigonellagenin.

Varshney and Sharma (1966) reported only diosgenin and gitogenin. Fazli (1967) reported, besides the forementioned sapogenins, the isolation from fenugreek seed of yamogenin, the 25β-epimer of diosgenin. He mentioned, also, a higher level of diene (50 percent of the weight of diosgenin).

Shortly afterwards, one more sapogenin, neogitogenin (VIII) (Figure 9.1) was isolated from Western Pakistan and Moroccan fenugreek seeds (Fazli and Hardman, 1971). A trace of tigogenin was detected by TLC from Moroccan seed only. Gitogenin was found only in the seed of both specimens. Trigonellagenin, previously mentioned by Soliman and Mustafa was considered to be a mixture of the major sapogenins, namely, diosgenin and yamogenin (Fazli and Hardman, 1971).

The total sapogenin content of the whole seed of fenugreek was 1.27 percent (25α-epimers, 62 percent and 25β-epimers, 38 percent) for the W. Pakistan seed and 1.50 percent (both epimers equal) for the Moroccan seed (Fazli and Hardman, 1971).

Dawidar and Fayez (1972) studied the sapogenin makeup of the plant at various stages of growth along with the different parts of the seeds and they revealed that the seedlings have the highest diosgenin (and other steroid sapogenin) content compared to all other stages of growth. Shortly afterwards, Dawidar *et al.* (1973) reinvestigated the fenugreek seeds grown in Egypt and reported for the first time the presence of neotigogenin (IV) (Figure 9.1).

Depending on the geographical source of the seed its sapogenin content, calculated as diosgenin, varied from 0.8–2.2 percent expressed on a moisture free basis (Fazli and Hardman, 1968). The highest sapogenin content was found in an Ethiopian sample and the lowest in a sample from Israel.

Fenugreek seed contains no free sapogenin but complex precursors, since frequently sapogenins occur in the plant as furostanol glycosides from which spirostanol glycosides are secondarily formed (Sauvaire and Baccou, 1978).

These glycosides (saponins) are limited to the fixed-oil containing embryo, but absent from the seed coats, namely the testa and the mucilage containing endosperm. The fenugreek seed is hard, flattened, brown to reddish-brown with a more or less parallel epipedal, without rounded edges. The widest surfaces are marked by a groove that divides the seed into two unequal parts. The smaller part contains the radicle, the larger part contains the cotyledons.

Saponins are not directly in association with the stored fat, but rather with the cell wall material and as free saponin in the circulatory system of the plant thus effecting easy transportation of the steroid and protecting the latter (Fazli and Hardman, 1971). Glycoside formation involving the cell wall (Blunden *et al.*, 1965; Hardman and Sofowora, 1971) may well be a method of steroid storage in the plant and of controlling excess steroid, thus preventing its interference in normal cellular mechanisms (Fazli and Hardman, 1971).

Sapogenins are released only after enzymic or acid hydrolysis (Blunden and Hardman, 1963).

The sapogenins available by the acid hydrolysis of fenugreek seeds are mainly the monohydroxysapogenins, diosgenin ([25 R]-spirost-5-en-3β-ol) and its (25S)-epimer yamogenin in a ratio of about 3 : 2. About 10 percent of their weight is a mixture of the two corresponding 5α-saturated monohydroxysapogenins, tigogenin and neotigogenin. In addition to these four sapogenins, there are very small percentages of each of their corresponding 2-hydroxy derivatives, namely yuccagenin (V) (Figure 9.1), lilagenin (VI) (Figure 9.1), gitogenin and neogitogenin, respectively (Cornish *et al.*, 1983). Sarsapogenin (IX) (Figure 9.1) and smilagenin (X) (Figure 9.1) were also isolated from the hydrolyzed seed. (Gupta *et al.*, 1986b) All these substances have a common cyclopentanoperhydro-phenanthrenic structure with twenty-seven carbon atoms and six rings.

Depending on the configuration of C_{25}, the 3, 26-biglycosides of the Δ^5-furostene type afford on hydrolysis diosgenin and yamogenin; the 5α-furostan type afford tigogenin and neotigogenin; the 5β-furostan type yield sarsapogenin or smilagenin, while the 2α, 5α-furostan type yield neogitogenin or gitogenin.

Also precursors of the type 3-peptide ester, 26-glucosides of Δ^5-furostene presumably exist from the evidence of the corresponding spirostene ester, fenugreekine (Ghosal *et al.*, 1974). On acid hydrolysis, it afforded diosgenin, yamogenin, (25R)-spirosta-3,5-diene, a mixture of three isomeric (2S,3R,4R-, 2S,3R,4S-, 2S,3S,4R-)-4-hydroxyisoleucine lactones (in a ratio of about 25 : 20 : 55, respectively), 4'-hydroxyisoleucine lactone and a C_{14}-dipeptide, which was partially characterized. Fenugreekine shows a number of interesting pharmacological activities (diuretic, cardiotonic, hypoglycemic, hypotensive, viristat against vaccinia virus and anti-inflammatory actions; Ghosal *et al.*, 1974; Che, 1991; Duke, 1992), which would account for the reported therapeutic uses of fenugreek in native medicine.

Fenugreek seeds mainly contain steroids of the 25S series, but during acid hydrolysis some of these are converted into the 25 R-spirostanes (Bogacheva *et al.*, 1976b).

The following furostanol glycosides have been isolated from the fenugreek seed: trigonelloside C (Figure 9.2) [(yamogenin) 3-0-α-L-rhamnopyranosyl(1 → 4) [α-L-rhamnopyranosyl (1 → 2)]-β-D-glucopyranoside 26-0-β-D-glucopyranoside] (Bogacheva *et al.*, 1976a, 1977a); its 22-0-methyl ether (Bogacheva *et al.*, 1977a); (neotigogenin) 3-0-α-L-rhamnopyranosyl (1 → 2) [β-D-glucopyranosyl (1 → 3)]-β-D-glucopyranoside 26-0-β-D-glucopyranoside, as its 22-0-methyl ether (Figure 9.2) (Hardman *et al.*, 1980); trigofoenosides A–G as their methyl ethers A1–G1 (Gupta *et al.*, 1984; 1985a,b; 1986a).

The structures of the original trigofoenosides have been determined as:

- (yamogenin) 3-0-α-L-rhamnopyranosyl(1 → 2)-β-D-glucopyranoside 26-0-β-D-glucopyranoside (A) (Figure 9.3) (Gupta *et al.*, 1985a);
- (neogitogenin) 3-0-α-L-rhamnopyranosyl(1 → 4)-β-D-glucopyranoside 26-0-β-D-glucopyranoside (B) (Figure 9.4) (Gupta *et al.*, 1986a);
- (gitogenin) 3-0-α-L-rhamnopyranosyl(1 → 4)-[α-L-rhamnopyranosyl(1 → 2)]-β-D-glucopyranoside 26-0-β-D-glucopyranoside (C) (Figure 9.4) (Gupta *et al.*, 1986a);
- (yamogenin) 3-0-α-L-rhamnopyranosyl(1 → 2)-[β-D-glucopyranosyl(1 → 3)]-β-D-glucopyranoside 26-0-β-D-glucopyranoside (D) (Figure 9.3) (Gupta *et al.*, 1985a);
- (tigogenin) 3-0-α-L-rhamnopyranosyl (1 → 2)-[β-D-xylopyranosyl(1 → 4)]-β-D-glucopyranoside 26-0-β-D-glucopyranoside (E) (Figure 9.4) (Gupta *et al.*, 1985b);
- (diosgenin) 3-0-α-L-rhamnopyranosyl (1 → 2)-β-D-glucopyranosyl (1 → 6)β-D-glucopyranoside 26-0-β-D-glucopyranoside (F) (Figure 9.3) (Gupta *et al.*, 1984); and
- (diosgenin) 3-0-α-L-rhamnopyranosyl(1 → 2)-[β-D-xylopyranosyl(1 → 4)]-β-D-glucopyranosyl(1 → 6)β-D-glucopyranoside 26-0-β-D-glucopyranoside (G) (Figure 9.3) (Gupta *et al.*, 1984).

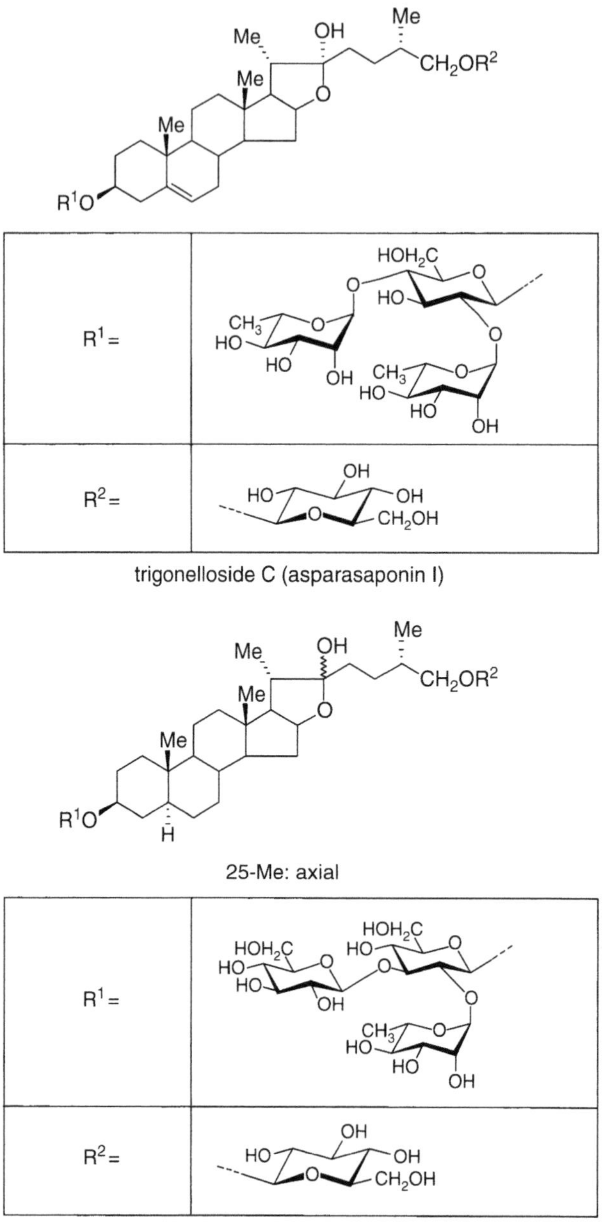

trigonelloside C (asparasaponin I)

Figure 9.2 Chemical structures of asparasaponin I and compound XII.

These furostanol glycosides appeared as a pair comprising the hydroxy- and methoxy-compounds. It has been observed that the furostanol glycosides when extracted with methanol undergo methylation yielding a mixture of 22-hydroxy and 22-methoxy derivatives (Tschesche *et al.*, 1972). In order to confirm that the 22-methoxy derivatives are probable artifacts, Gupta *et al.* (1984) studied a separate extraction with pyridine and found that the 22-methoxy compounds were completely absent.

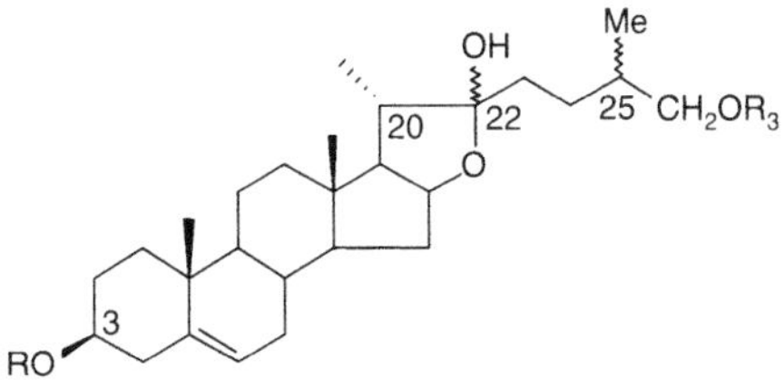

The steroid/furostanol core structure with labels: OH, Me, 20, 22, 25, CH₂OR₃, O, RO, 3.

R	R₁	R₂	R₃	
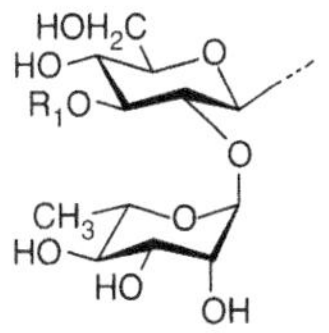 HOH₂C, HO, R₁O, O, CH₃, HO, HO, OH	H		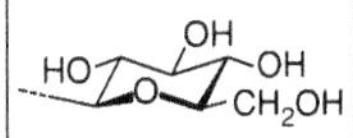 HO, OH, OH, O, CH₂OH	trigofoenoside A 25-Me: axial
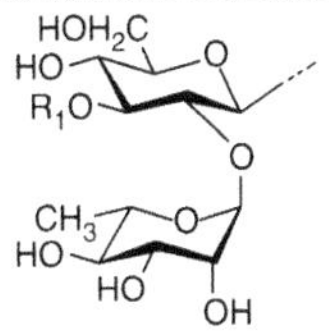 HOH₂C, HO, R₁O, O, CH₃, HO, HO, OH	β-D-glucopyranoside		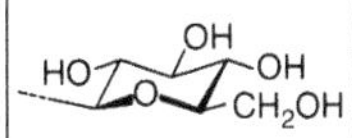 HO, OH, OH, O, CH₂OH	trigofoenoside D 25-Me: axial
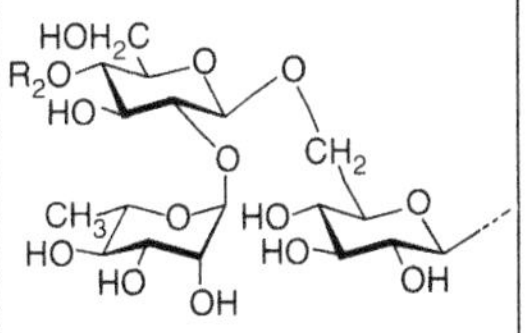 HOH₂C, R₂O, HO, O, CH₂, CH₃, O, HO, HO, HO, OH, O, OH		H	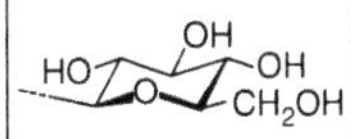 HO, OH, OH, O, CH₂OH	trigofoenoside F 25-Me: equatorial
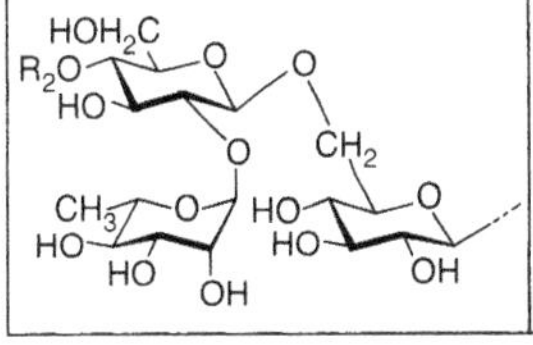 HOH₂C, R₂O, HO, O, CH₂, CH₃, O, HO, HO, HO, OH, O, OH	β-D-xylopyranoside		HO, OH, OH, O, CH₂OH	trigofoenoside G 25-Me: equatorial

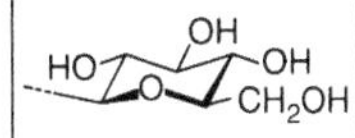

Figure 9.3 Chemical structures of trigofoenosides A, D, F, G.

Six furostanol glycosides called trigoneosides Ia, Ib, IIa, IIb, IIIa, IIIb were isolated from fenugreek seed originating from India, together with two known saponins, trigofoenoside A and its 25-R epimer, glycoside D (Yoshikawa *et al.*, 1997). Their structures were determined as:

– 26-O-β-D-glucopyranosyl-(25S)-5α-furostane-2α,3β,22ξ,26-tetraol 3-O-[(β-D-xylopyranosyl)(1 → 6)]-β-D-glucopyranoside (trigofoenoside Ia) (Figure 9.5); and its 25R-epimer (trigofoenoside Ib) (Figure 9.6);

– 26-O-β-D-glucopyranosyl-(25S)-5β-furostane-3β,22ξ,26-triol 3-O-[(β-D-xylopyranosyl)(1 → 6)]-β-D-glucopyranoside (trigofoenoside IIa) (Figure 9.5); and its 25R-epimer (trigofoenoside IIb) (Figure 9.6);

– 26-O-β-D-glucopyranosyl-(25S)-5α-furostane-3β,22ξ,26-triol 3-O-[(α-L-rhamnopyranosyl)(1 → 2)]-β-D-glucopyranoside (trigofoenoside IIIa) (Figure 9.5); and its 25R-epimer (trigofoenoside IIIb) (Figure 9.6).

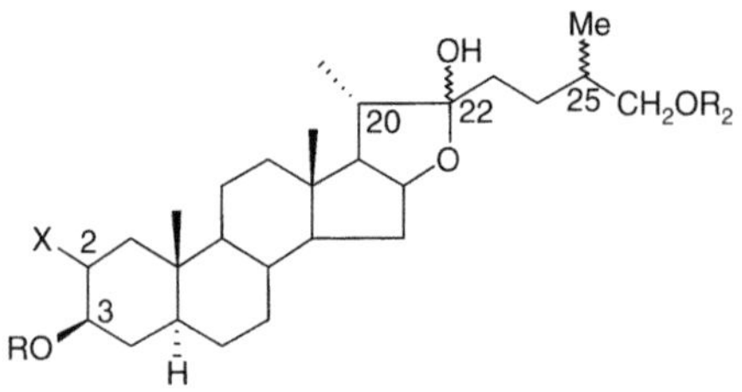

X	R	R₁	R₂	
OH	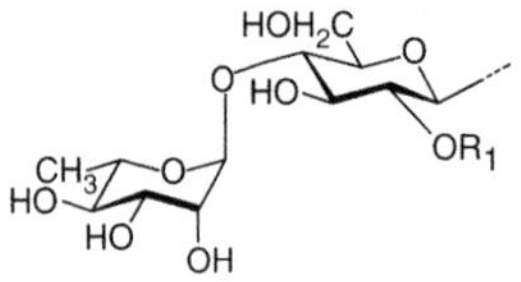 (sugar structure with HOH₂C, HO, CH₃, HO, OH, OR₁)	H	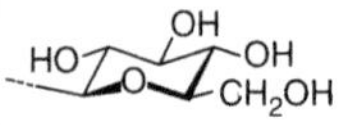 (sugar: HO, OH, OH, CH₂OH)	trigofoenoside B 25-Me: axial
OH	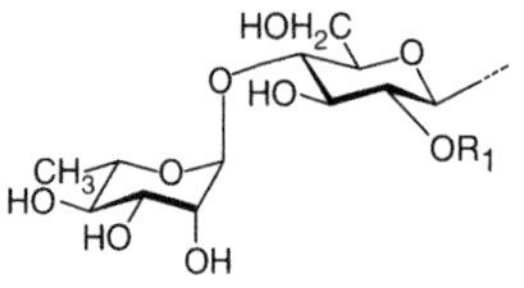 (sugar structure with HOH₂C, HO, CH₃, HO, OH, OR₁)	α-L-rhamnopyranoside	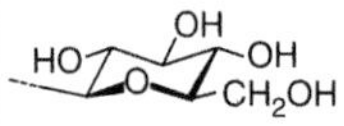 (sugar: HO, OH, OH, CH₂OH)	trigofoenoside C 25-Me: equatorial
H	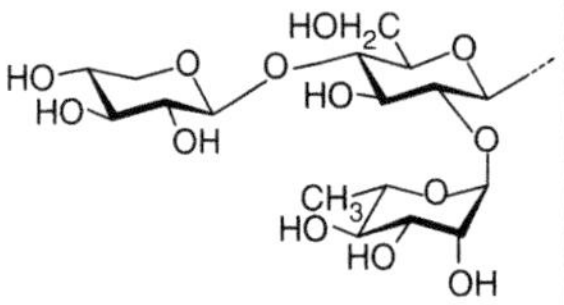 (sugar structure with HO, HO, OH, HOH₂C, HO, CH₃, HO, OH)		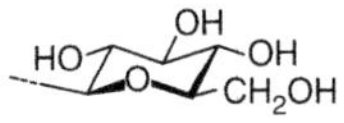 (sugar: HO, OH, OH, CH₂OH)	trigofoenoside E 25-Me: equatorial

Figure 9.4 Chemical structures of trigofoenosides B, C, E.

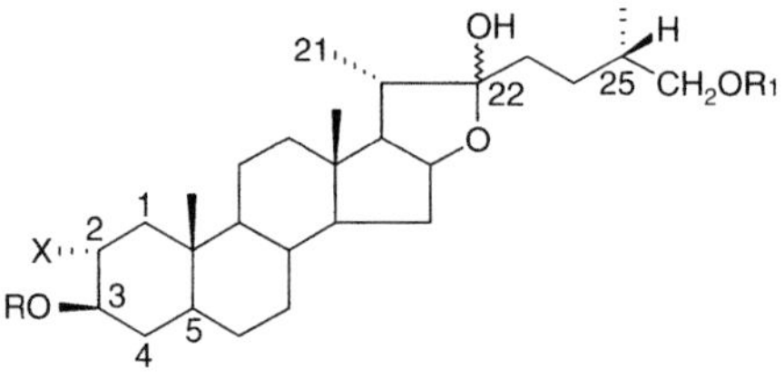

X	R	R₁	
OH	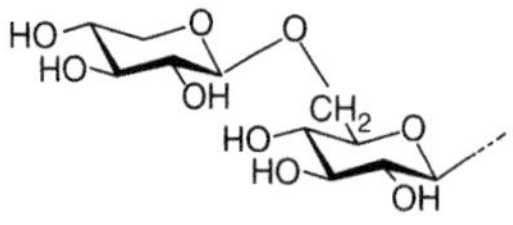 (sugar structure with HO, HO, OH, CH₂, HO, HO, OH)	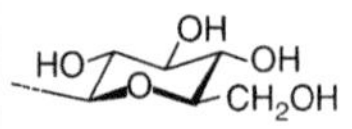 (sugar: HO, OH, OH, CH₂OH)	trigoneoside Ia
H	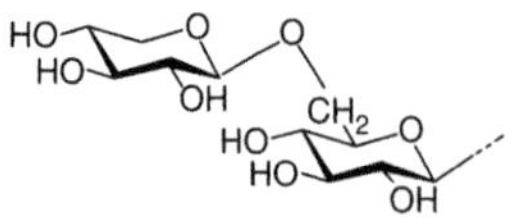 (sugar structure with HO, HO, OH, CH₂, HO, HO, OH)	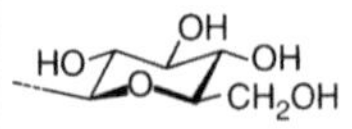 (sugar: HO, OH, OH, CH₂OH)	trigoneoside IIa
H	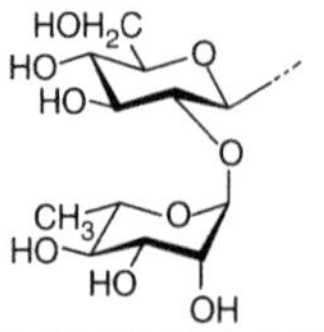 (sugar structure with HOH₂C, HO, HO, CH₃, HO, HO, OH)	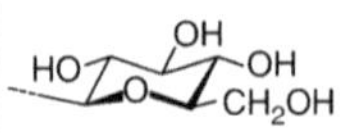 (sugar: HO, OH, OH, CH₂OH)	trigoneoside IIIa

Figure 9.5 Chemical structures of trigoneosides Ia, IIa, IIIa.

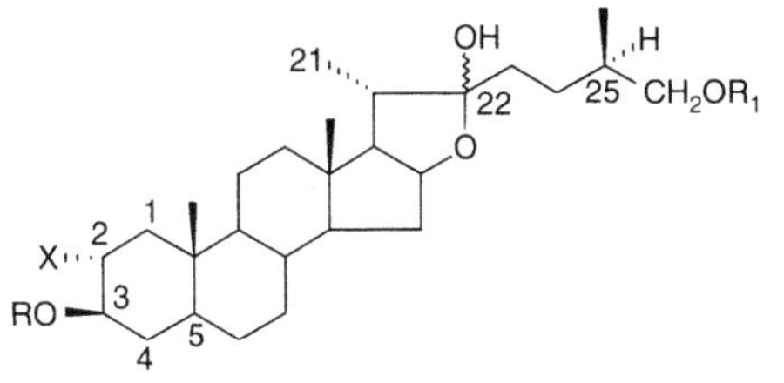

X	R	R1	
OH	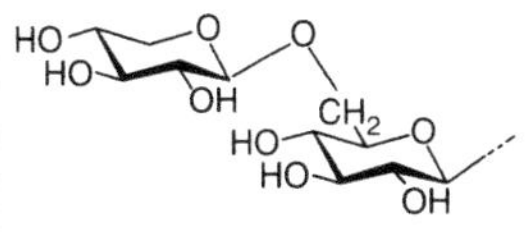	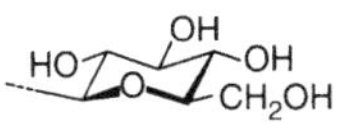	trigoneoside Ib
H	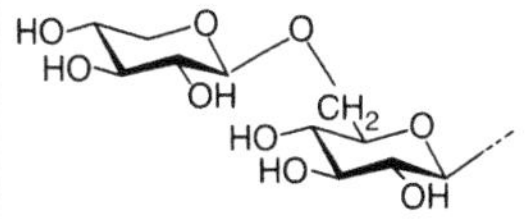	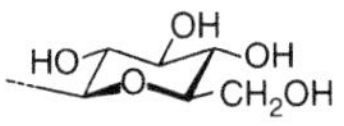	trigoneoside IIb
H	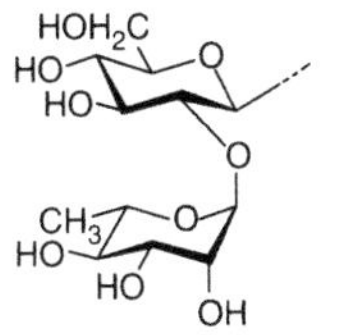	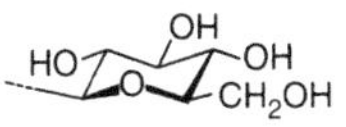	trigoneoside IIIb

Figure 9.6 Chemical structures of trigoneosides Ib, IIb, IIIb.

Acid hydrolysis of trigoneosides Ia–IIIa furnished the (25S)-aglycones neogitogenin, sarsapogenin and neotigogenin, while acid hydrolysis of trigoneosides Ib–IIIb furnished their 25R-epimers, namely, gitogenin, smilagenin and tigogenin, respectively (Yoshikawa *et al.*, 1997).

Further investigation of the Indian fenugreek seeds led to the isolation of seven new furostanol saponins, called trigoneosides IVa, Va, Vb, VI, VIIb, VIIIb, IX along with the known furostanol saponins, compound C, glycoside F (Figure 9.9) and trigonelloside C (Figure 9.2). The structures of six of these furostanol saponins were assigned as follows (Yoshikawa *et al.*, 1998):

- 26-*O*-β-D-glucopyranosyl-(25S)-furost-5-ene-3β,22ξ,26-triol 3-*O*-[α-L-rhamnopyranosyl (1→2)][β-D-glucopyranosyl (1→4)]-β-D-glucopyranoside (trigoneoside IVa) (Figure 9.7);
- 26-*O*-β-D-glucopyranosyl-(25S)-furost-5-ene-3β,22ξ,26-triol 3-*O*-{α-L-rhamnopyranosyl (1→2)} {[β-D-xylopyranosyl (1→4)] [β-D-glucopyranosyl (1→6)]-β-D-glucopyranosyl (1→3)-β-D-glucopyranosyl (1→4)}-β-D-glucopyranoside (trigoneoside Va) (Figure 9.7); and its 25R-epimer (trigoneoside Vb) (Figure 9.7);
- 26-*O*-β-D-glucopyranosyl-furost-5, 25(27)-diene-3β,22ξ,26-triol 3-*O*-{α-L-rhamnopyranosyl(1→2)} {[β-D-xylopyranosyl (1→4)] [β-D-glucopyranosyl (1→6)]-β-D-glucopyranosyl (1→3)-β-D-glucopyranosyl (1→4)]}-β-D-glucopyranoside (trigoneoside VI) (Figure 9.7);
- 26-*O*-β-D-glucopyranosyl-(25R)-furost-5-ene-3β,22ξ,26-triol 3-*O*-{α-L-rhamnopyranosyl)(1→2)} {[β-D-xylopyranosyl (1→4)] [β-D-xylopyranosyl (1→6)]-β-D-glucopyranosyl (1→3)-β-D-glucopyranosyl (1→4)}-β-D-glucopyranoside (trigoneoside VIIb) (Figure 9.7);

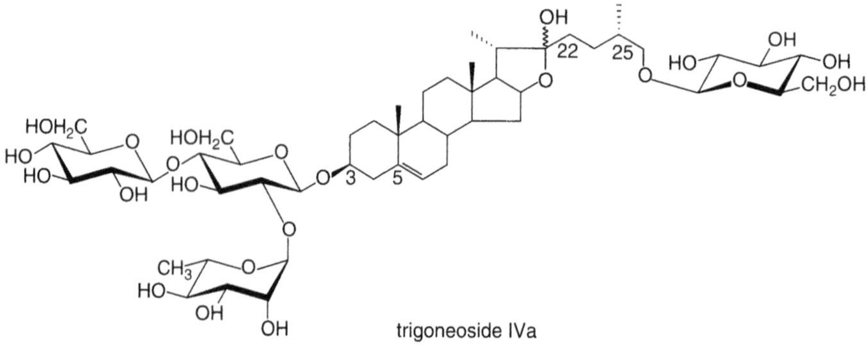

trigoneoside IVa

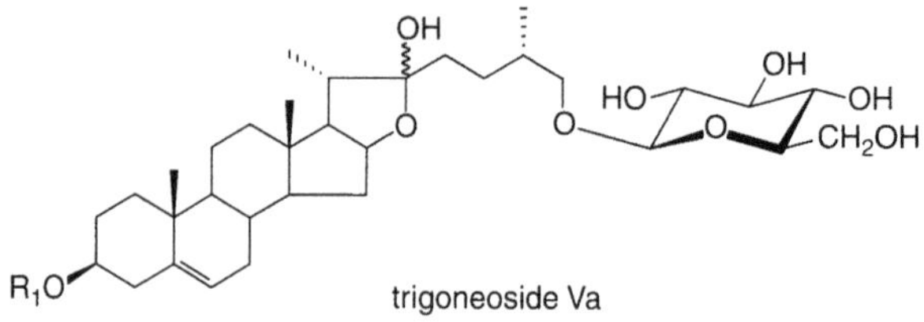

trigoneoside Va

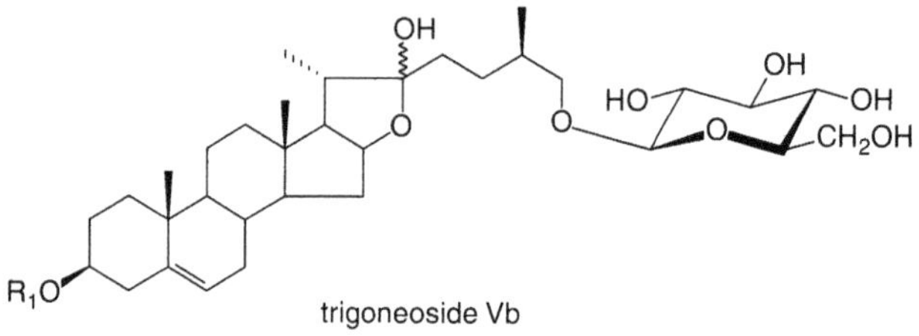

trigoneoside Vb

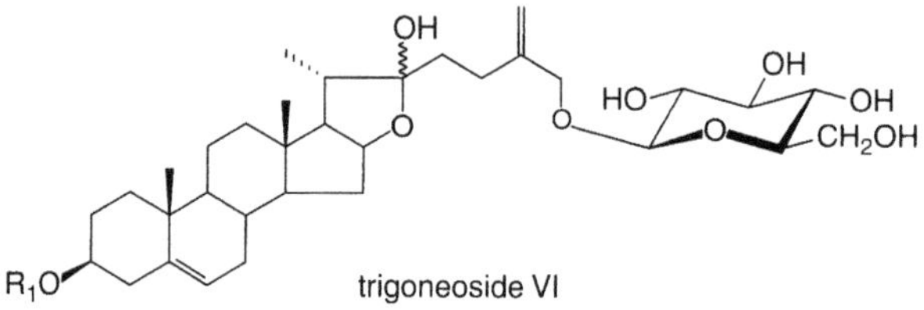

trigoneoside VI

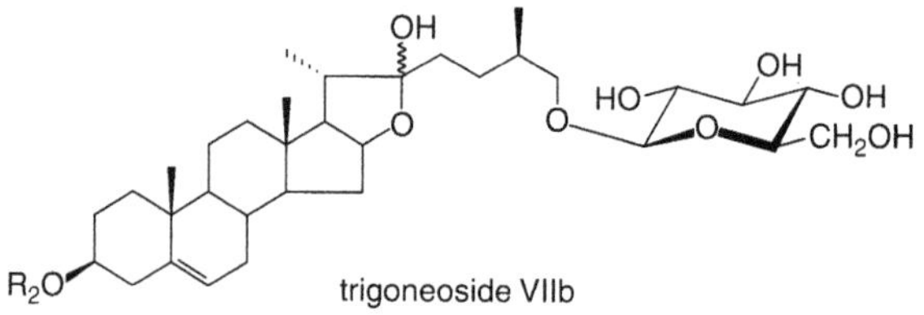

trigoneoside VIIb

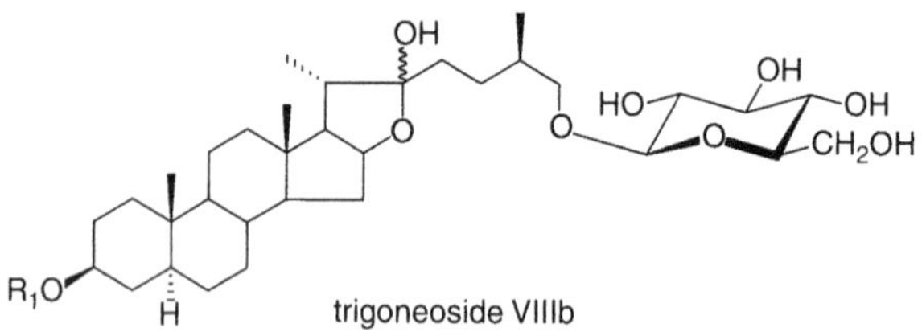

trigoneoside VIIIb

Figure 9.7 (Continued)

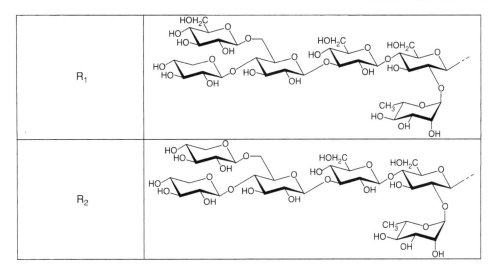

R₁	
R₂	

Figure 9.7 Chemical structures of trigoneosides IVa, Va, Vb, VI, VIIb, VIIIb.

– 26-O-β-D-glucopyranosyl-(25R)-5α-furostane-3β,22ξ,26-triol 3-O-{α-L-rhamnopyranosyl (1→2)} {[β-D-xylopyranosyl (1→4)] [β-D-glucopyranosyl (1→6)]-β-D-glucopyranosyl (1→3)-β-D-glucopyranosyl (1→4)}-β-D-glucopyranoside (trigoneoside VIIIb) (Figure 9.7).

The structure of trigoneoside IX has not yet been elucidated.

Recently, six new furastanol glycosides called trigoneosides Xa, Xb, XIb, XIIa, XIIb and XIIIa were isolated from the seeds of the Egyptian *T. foenum-graecum* L. together with the six known furastanol-type steroid saponins: trigoneosides Ia (Figure 9.5), Ib (Figure 9.6) and Va (Figure 9.7), trigonelloside C (Figure 9.2), glycoside D (Figure 9.9) and compound C (Figure 9.9) (Murakami *et al.*, 2000).

The structures of the new furastanol glycosides were determined as:

– 26-O-β-D-glucopyranosyl-(25S)-5α-furostane-2α,3β,22ξ,26-tetraol 3-O-α-L-rhamnopyranosyl (1→2)-β-D-glucopyranoside (trigoneoside Xa) (Figure 9.8); and its 25R-epimer (trigoneoside Xb) (Figure 9.8);

– 26-O-β-D-glucopyranosyl-(25R)-5α-furostane-2α,3β,22ξ,26-tetraol 3-O-β-D-xylopyranosyl (1→4)-β-D-glucopyranoside (trigoneoside XIb) (Figure 9.8);

– 26-O-β-D-glucopyranosyl-(25S)-furost-4-ene-3β,22ξ,26-triol 3-O-α-L-rhamnopyranosyl (1→2)-β-D-glucopyranoside (trigoneoside XIIa) (Figure 9.8); and its 25R-epimer (trigoneoside XIIb) (Figure 9.8);

– 26-O-β-D-glucopyranosyl-(25S)-furost-5-ene-3β,22ξ,26-triol 3-O-α-L-rhamnopyranosyl (1→2)-[β-D-glucopyranosyl (1→3)-β-D-glucopyranosyl (1→4)]-β-D-glucopyranoside (trigoneoside XIIIa) (Figure 9.8).

Seven spirostanol saponins have also been isolated from the fenugreek seeds, which were named graecunins H–N. All are glycosides of diosgenin with different sugar moieties. Graecunins H, I, J and K contain varying amounts of glucose and rhamnose, whereas graecunin-N contains glucose, arabinose, xylose and rhamnose. Partial structures were assigned to some of these glycosides (Varshney and Begs, 1978).

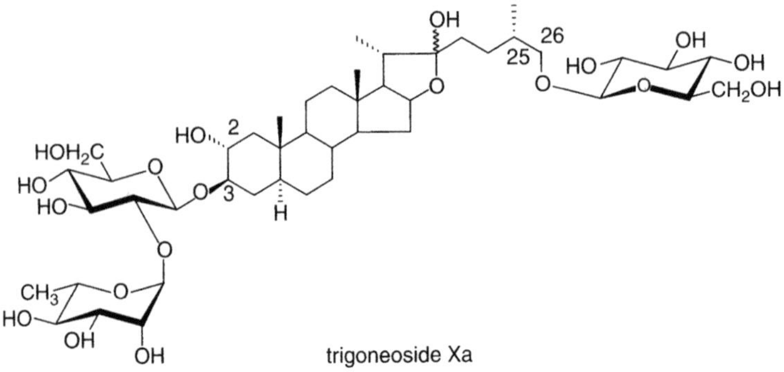

trigoneoside Xa

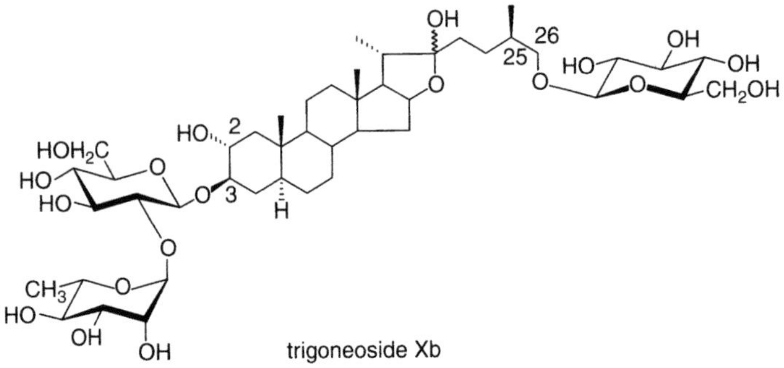

trigoneoside Xb

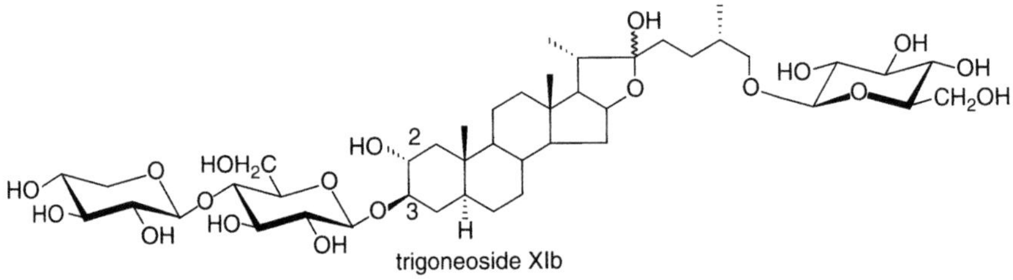

trigoneoside XIb

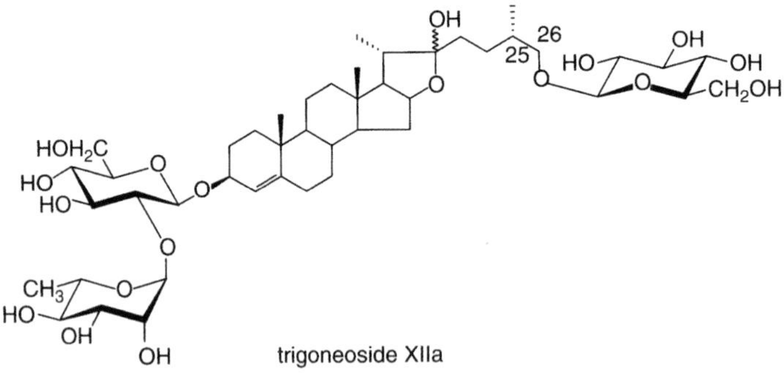

trigoneoside XIIa

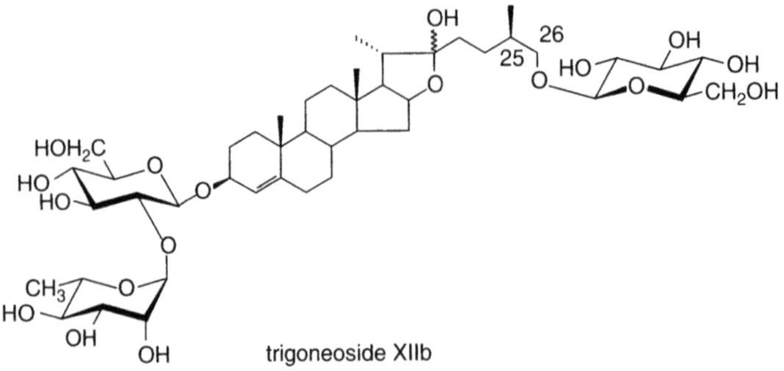

trigoneoside XIIb

Figure 9.8 (Continued)

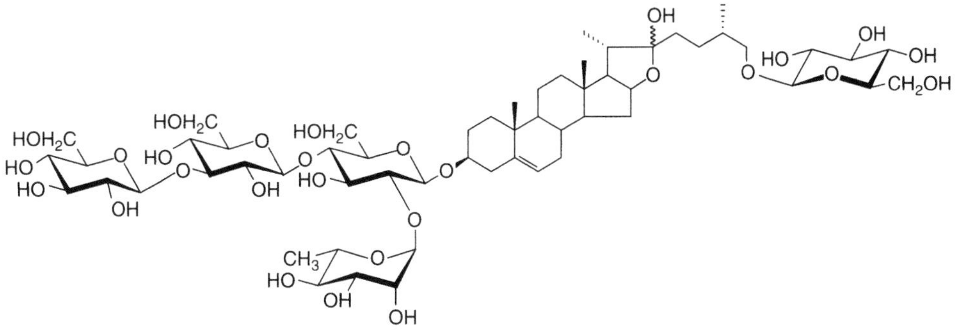

trigoneoside XIIIa

Figure 9.8 Chemical structures of trigoneosides Xa, Xb, XIb, XIIa, XIIb, XIIIa.

A saponin, named fenugrin B, was also obtained from the fenugreek seed. This compound, on acid hydrolysis, gave diosgenin and the sugars: glucose, arabinose and rhamnose (Gangrade and Kaushal, 1979).

From fenugreek leaves five spirostanol saponins have been isolated and named graecunin-B, -C, -D, -E and -G. Two trace compounds, named graecunin-A and -F, were also isolated in too small amounts to characterize them (Varshney and Jain, 1979; Varshney *et al.*, 1984).

Graecunin-E and graecunin-G have been shown to be (diosgenin) 3-0-α-D-glucopyranosyl $(1\rightarrow 4)$ α-L-rhamnosyl $(1\rightarrow 2)$ α-L-rhamnosyl $(1\rightarrow 6)$ α-D-glucopyranoside and (diosgenin) 3-0-α-D-glucopyranosyl $(1\rightarrow 2)$ α-L-rhamnopyranosyl $(1\rightarrow 6)$ α-D-glucopyranoside, respectively (Varshney *et al.*, 1984). Partial structures were assigned to the other glycosides. Graecunin B (Varshney *et al.*, 1977; Varshney and Jain, 1979; Varshney *et al.*, 1984) and D (Varshney and Jain, 1979; Varshney *et al.*, 1984) contained glucose, xylose and rhamnose in the molar ratio $4:1:2$ and $4:1:1$, respectively. Graecunin C (Varshney and Jain, 1979; Varshney *et al.*, 1984) contained glucose and rhamnose in the molar ratio $4:1$.

Oils obtained by the separate extraction of the powdered dried leaf, stem and root from Moroccan plants yielded, after saponification, squalane-like hydrocarbons and β-sitosterol, but no free sapogenin or spirostadiene. Acid hydrolysis of the defatted powdered leaf yielded 25α- and 25β-spirosta-3,5-diene and a 1:1 mixture of diosgenin and yamogenin. Stem and root, when similarly treated, showed the same steroids in trace amounts. Gitogenin was not detected in the leaf, stem or root (found only in the seed) (Fazli and Hardman, 1971).

The increase in the yield of steroidal sapogenin from fresh and dried plant material on its incubation with water under defined conditions has been reported (Blunden and Hardman, 1963; Blunden and Hardman, 1965; Hardman and Brain, 1971; Hardman and Sofowora, 1971; Hardman and Wood, 1971a; Hardman and Wood, 1971b). Blunden *et al.* (1965) showed that large increases in sapogenin yield could be obtained by incubating harvested plant material from various species and morphological parts in an excess of water. The process was enzymic and the endogenous enzymes could be replaced, at least partly, by cell wall degrading enzymes. The phenomenon occurs irrespective of the nature of the sapogenin, the nature of the tissue and of the plant genus.

The work has been extended to the sapogenin yielding capacity of the fenugreek seed, when the endogenous enzymes are allowed to function alone or in the presence of an additive. Aqueous incubation at 37°C (tropical temperature) prior to acid hydrolysis resulted in an increase of only 10 percent on average, which could be attributed to the release of the sapogenin by the enhanced activation of the endogenous enzyme system of the seed. The addition of mevalonate or choles-terol did not result in an increase in sapogenin (Hardman and Fazli, 1972b).

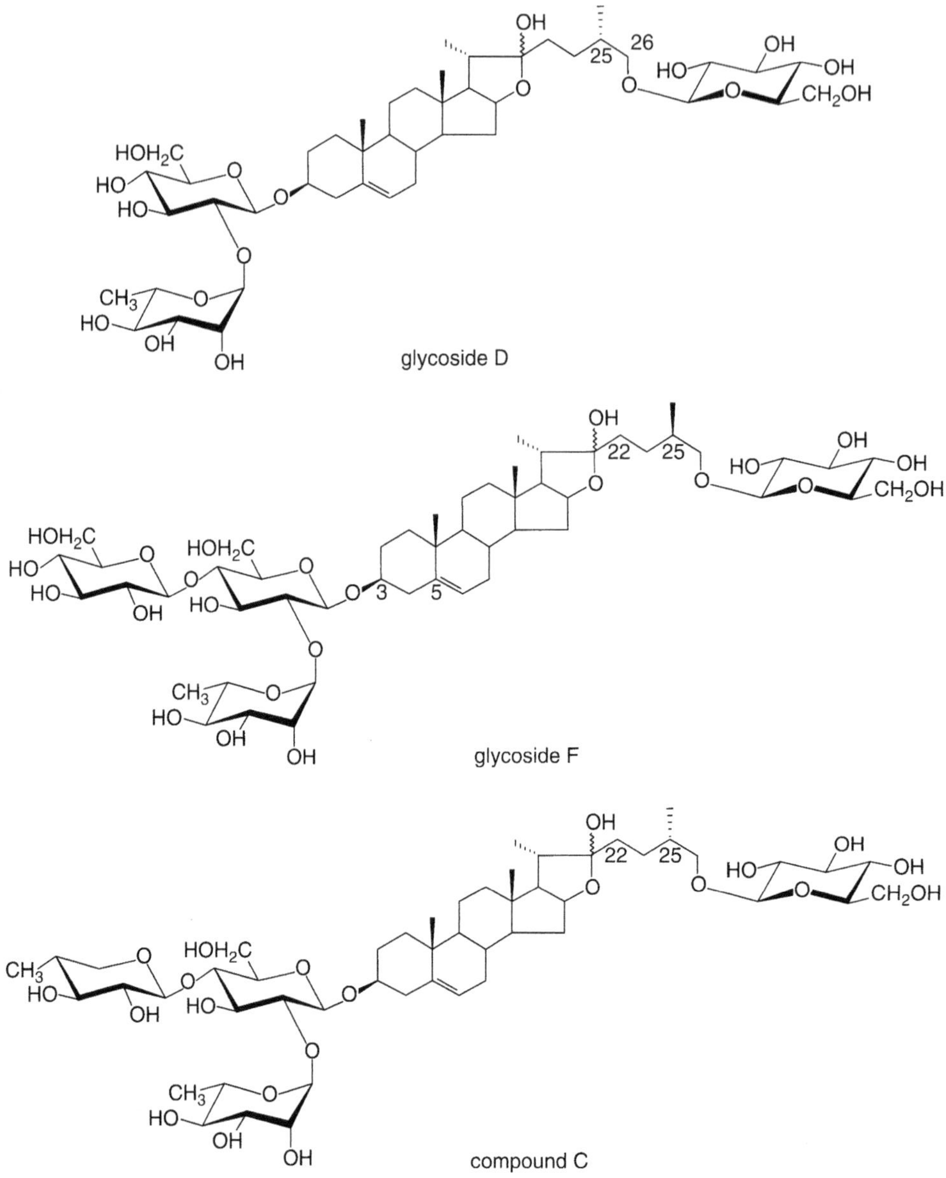

Figure 9.9 Chemical structures of glycoside D, glycoside F and compound C.

The optimal incubation conditions prior to acid hydrolysis for a high yield of sapogenin (0.90 percent) of fenugreek whole seed were estimated to be: temperature 45°C, initial pH 4.0 with aeration and shaking for 4 days (Elujoba and Hardman, 1985a). If the seeds are ground, the yields decrease to 0.50 percent. This reduced yield could be due to the increased binding of sapogenins in the seed during the grinding process. The seed constituents (e.g. aminoacids, proteins, mucilage, etc.), which otherwise are separately located in the seed, during grinding come in close contact with furostanol glycosides thus resulting in the additional production of acid-resistant "bound" forms of sapogenin. A higher yield (up to 1.65 percent) is obtained if the ground seeds are incubated with enzymes (Elujoba and Hardman, 1985b).

Hardman and Brain (1971) reported that incubation of the whole seed of *T. foenum-graecum* L. with their synthetic or natural plant growth regulators increased the sapogenin yield by up to 35 percent. The process is concentration and time dependent. The variation in the steroid levels and distribution of these compounds with alteration in medium composition and culture age have been investigated in tissue cultures (Brain and Lockwood, 1976; Lockwood and Brain, 1976; Hardman and Stevens, 1978). It can be concluded that the nature of the growth hormone produces significant differences in the yield of monohydroxysapogenin and individual sterols. Since cholesterol, or a closely related compound, has been implicated in sapogenin biosynthesis, one cholesterol-blocking agent, such as 2-(*p*-chloro-phenoxy) 2-methyl-propionic acid ethyl ester, was incubated with the whole fenugreek seed up to 24 h. The subsequent observed increase was about 20 percent. The phenoxyacetic acids are thought to act like the natural plant auxin indole-3-acetic acid (IAA), possibly by prevention of the destruction of the endogenous hormone. Incubation with fenugreek gave rises in the total sapogenin of about 35 percent after 24 h.

Hardman and Brain (1972) studied the variation in the yield of total and individual 25α- and 25β-sapogenins on storage of whole fenugreek seed. Total sapogenin yield under experimental conditions (except for samples stored at 5°C) showed a decline over about 50 days, followed by a rise and fall. The initial decline in the total sapogenin was due in all cases to a selective loss of the 25α-form.

Sauvaire and Baccou (1978) investigated the conditions (nature and concentration of the acid and solvents, as well as the ratio between quantity of substrate and volume of hydrolyzing solution) for an efficient acid hydrolysis of steroidal glycosides resulting in a high yield of diosgenin and avoiding formation of spirosta-3,5-diene.

The following methods of detection for steroidal sapogenins in plant material provided from the genus *Trigonella* were applied: blood hemolysis, color reaction, infrared spectrophotometric assay and thin layer chromatography (Fazli and Hardman, 1968; Hardman and Jefferies, 1971; Hardman and Fazli, 1972a; Dawidar and Fayez, 1972).

A rapid quantitative determination of C_{25} epimers in plants was described as both occur in plant tissue (depending on their ratio from a number of factors, e.g. morphological part, stage of development etc.). Prior to an *in situ* hydrolysis of the saponin by aqueous hydrolytic acid and a chloroform extraction, the measurement of the specific spirostan absorption and calculation of the absorbance of the bands at $915\,cm^{-1}$ and $900\,cm^{-1}$ enables the determination of the 25β- and 25α- forms separately, with a 3–10 percent overall error for individual C_{25} epimers and 3–5 percent for total sapogenin (Brain *et al.*, 1968). The IR spectrophotometric analysis of crude extracts was later shown by Hardman and Jefferies (1972) to give high values and replaced by column chromatography preceding IR analysis. The method removes sterols, steryl esters, spirostadienes and dihydroxysapogenins, such as gitogenin (not useful as a raw material) from the fraction containing diosgenin and yamogenin and it has been further improved (Jefferies and Hardman, 1976).

Gas–liquid chromatography has been proposed (Knight, 1977) for the analysis of fenugreek sapogenins (as trimethylsilyl ethers and trifluoroacetates). The method has the possibility to separate the C_{25} epimers from each other and from their 5α-dihydro analogs and the more polar $2\alpha,3\beta$-dihydroxy-steroids.

Jain and Agrawal (1987) studied the effect of physical (UV and γ-irradiation) and chemical mutagens (ethyl methane sulphonate (EMS), methyl methane sulphonate (MMS) and sodium azide (NaN_3)) in tissue culture. A two- to four-fold increase in the sapogenin content was observed in the plants and seeds obtained from fenugreek seeds treated with a low concentration of the chemical mutagens and an approximately two-fold increase was observed with UV (2 h irradiation), while γ-irradiation could enhance the yield by *c.* 85 percent only (Jain and Agrawal, 1994).

The crude saponins of fenugreek seed showed a hypocholesterolemic activity in experiments (Sharma, 1986).

Sterols are present in all parts of the plant and occured in both a combined and free state (Fazli and Hardman, 1971). β-Sitosterol was found in leaf extracts of Indian samples (Sood, 1975); β-sitosterol-D-0-glucoside (=daucosterol) was isolated from the whole plant (Parmar *et al.*, 1982).

Small amounts of cholesterol and two others sterols, not identified, were detected. Cholesterol is the main sterol involved in the biosynthesis of steroidal sapogenins (Hardman and Fazli, 1972b). Lower incorporation of cholesterol into gitogenin than into diosgenin suggests that gitogenin may be formed from diosgenin. Such a conversion is also supported by the finding that in a growing plant gitogenin, in contrast to diosgenin, is absent from the leaves and is found only in the ripe seed and young seedlings (Hardman and Fazli, 1972c).

Khanna and Jain (1973) reported for the first time the production and isolation of sterols and sapogenins from static cultures of fenugreek. Higher levels of β-sitosterol, stigmasterol and of the steroidal sapogenin were obtained in 8 week old static cultures compared to that of seeds.

The amounts of stigmasterol, campesterol, β-sitosterol and cholesterol and the ratio of stigmasterol to sitosterol in the free and the bound sterol fraction from static cultures were measured by GLC analysis of their TMS ethers (Hardman and Stevens, 1978).

Furthermore, the sterolic composition of the plant has been reinvestigated. It is characterized by a quasi absence of stigmasterol and by the presence of Δ^7-sterols and of an unusual sterol, pollinastanol (14α-methyl-9β,19-cyclo-5α-cholestan-3β-ol) (Brenac and Sauvaire, 1996).

Recently, from the ethanol extract of the seeds, six triterpenoids were isolated and identified as lupeol, 31-*nor*cycloartanol, betulin, betulinic acid, soyasaponin I and soyasaponin I methyl ester (Shang *et al.*, 1998).

Other *Trigonella* species[1]

Besides *T. foenum-graecum* L., seeds from *T. coerulea* (L.) Ser., T. *corniculata* L., *T. cretica* (L.) Boiss. contain different amounts of various steroidal sapogenins with diosgenin being predominant. Steroidal sapogenins were absent from *T. calliceras* Fisch. ex Bieb. By the blood analysis test *T. monspeliaca* L. and *T. polycerata* L., and by the color reaction method *T. hamosa* L. and *T. polycerata* L. gave positive results. *T. platycarpa* L. and *T. radiata* Boiss. gave negative results in both tests (Hardman and Fazli, 1972a).

Bohannon *et al.* (1974) examined the seeds from twenty-seven species of *Trigonella* for sapogenin, but none was richer than *T. foenum-graecum* L. in the component calculated as diosgenin, but presumably also containing yamogenin and tigogenin. In addition to *T. foenum-graecum* L. only five species contain at least 0.2 percent diosgenin and analog substances: *T. coerulea* (L.) Ser., *T. corniculata* (L.) L., *T. fischeriana* Ser., *T. gladiata* Stev. and *T. sibthorpii* Boiss.

The following *Trigonella* species contain less than 0.2 percent diosgenin, usually less than 0.1 percent: *T. anguina* Del., *T. arabica* Del., *T. arcuata* C.A. Mey., *T. brachycarpa* (Fisch.) Moris, *T. caelesyriaca* Boiss., *T. calliceras* Fisch., *T. cretica* (L.) Boiss., *T. emodi* Benth., *T. incisa* Benth., *T. kotschyi* Fenzl. ex Boiss., *T. monantha* C.A. Mey. *T. monspeliaca* L., *T. noëana* Boiss., *T. orthoceras* Kar. and Kir., *T. polycerata* L., *T. rigida* Boiss. and Bal., *T. spicata* Sibth & Sm., *T. stellata* Forssk., *T. suavissima* Lindl., *T. uncata* Boiss. and Noe [= *T. glabra* subsp. *uncata* (Boiss. and Noe) Lassen].

Diosgenin and 25α-spirosta-3, 5-diene were detected in roots, stem, leaves and pericarp of *T. maritima* Poiret and *T. stellata* Forssk., while only gitogenin was found in the seeds of both plants (Balbaa *et al.*, 1977).

---

1 The botanical names have been completed according to the Index Kewensis (Hooker and Jackson 1960).

Medicagenic acid, a triterpene sapogenin of quite limited occurrence, was detected in the seeds of the following species: *T. geminiflora* Bunge, *T. monpseliaca* L., *T. noëana* Boiss. and *T. polycerata* L. (Jurzysta *et al.*, 1988).

The composition of sterols was investigated in the following species: *T. foenum -graecum* L. (See p. 145), *T. calliceras* Fisch. ex Bieb., *T. corniculata* (L.) L. (see below), *T. caerulea* (L.) Ser. (see below), *T. melilotus caeruleus*[2] (L.) Ascherson and Graebner, *T. cretica* (L.) Boiss. and *T. monspeliaca* L.

Sitosterol and 24-methyl-cholesterol are the main sterols in all species except in *T. monspeliaca* L. Stigmasterol, usually well represented in plants, shows a low level in fenugreek and *T. cretica* (L.) Boiss. Pollinastanol was absent in *T. calliceras* Fisch. ex Bieb. and *T. monspeliaca* L., but present in all the other species, with higher levels in *T. caerulea* (L.) Ser. and *T. melilotus-caeruleus*[2] (L.) Ascherson and Graebner. These last two species also present highly similar compositions. By contrast, *T. monspeliaca* L. shows a composition very different from the other species. In complement of the absence of pollinastanol and the very low levels of sitosterol and 24-methyl-cholesterol, this species also presents high contents in α-spinasterol (absent in all the other species) and Δ^7-stigmastenol (only present in fenugreek but at a very low percentage) (Brenac and Sauvaire, 1996).

T. caerulea (L.) Ser. (=*T. coerulea* (L.) Ser.)

Diosgenin was extracted from seed and tissue cultures (Zambo and Szilagyi, 1982).

Glycosides of furost-5-en-3β,22,26-triol connected with the sugars glucose, rhamnose and xylose in different orders of bonding were isolated from the seeds (Kogan and Bogacheva, 1978). By incubating ground seeds with protosubtilin (*Bacillus subtilis* proteinase) for conversion of the contained furostanol glycosides to spirostan, yields of the genins during subsequent acid hydrolysis were increased. (25S)-Spirostadiene, diosgenin, gitogenin and its 25S-epimer neogitogenin, but not tigogenin, neotigogenin or yamogenin, were obtained (Bogacheva *et al.*, 1976c). Methanol extract of the seeds yielded the 22-methyl ether of protodioscin, assigned as 3β-(α-L-rhamnopyranosyl) (1→4)-α-L-rhamnopyranosyl-(1→2)-(β-D-glucopyranosyloxy-26-(β-D-glucopyranosyloxy)-22-α-methoxy-25R)-furost-5-en (Bogacheva *et al.*, 1977b).

The sterolic composition of the seeds is characterized by high levels of sitosterol, stigmasterol and 24-methyl cholesterol with lower amounts of cholesterol, pollinastanol and Δ^5-avenasterol. Small amounts of Δ^7-cholesterol, 24-methylene-cholesterol, Δ^7-campesterol, stigmastanol and fucosterol were detected (Brenac and Sauvaire, 1996).

T. corniculata L. (=*T. balansae* Boiss. and Reut.)

Varshney and Sood (1969) have reported the predominant sapogenin of the seeds to be the dihydroxysapogenin, yuccagenin (2α,3β-dihydroxy-25α-spirost-5-ene) being 70 percent, and diosgenin 25 percent of the total genins.

Diosgenin was found in the seed (Hardman and Fazli, 1972a; Bohannon *et al.*, 1974). The diosgenin plus yamogenin content was estimated to be about 0.15 percent on a moisture free basis (ratio of diosgenin to yamogenin 3 : 1) (Puri *et al.*, 1976).

Flowers were found to contain diosgenin, tigogenin and gitogenin (in a ratio of 70 : 15 : 5), while the leaves contain diosgenin as the main compound and some tigogenin (Varshney and Sood, 1971).

An increase in diosgenin and tigogenin levels was observed in the plants obtained from seeds treated with low concentrations of mutagens (Jain and Agrawal, 1987). Attempts were made to regulate the synthesis of diosgenin by induced mutagenesis (Mahna *et al.*, 1994).

---

2 It has been fused to *T. coerulea* (L.) Ser.

β-Sitosterol was isolated from seeds (Atal and Sood, 1964). Cholesterol, Δ^7-cholesterol, pollinastanol, 24-methyl- and 24-methylene-cholesterol, stigmasterol, Δ^7-campesterol, stigmastanol, fucosterol, Δ^5-avenasterol were also detected and their percentage in total sterols was estimated (Brenac and Sauvaire, 1996).

T. occulta Ser.

The seeds were found to contain as much as 0.32 percent of diosgenin with appreciably lower concentrations of the other two sapogenins, gitogenin (0.04 percent) and tigogenin (0.01 per cent) (Jain, 1976a). Hydrolyzed tissue culture yielded, besides the above mentioned sapogenins, β-sitosterol and traces of stigmasterol (Jain *et al.*, 1977). β-Sitosterol was also isolated from seeds (Jain, 1976b).

N-compounds

Trigonella foenum-graecum L.

Trigonelline (Karrer, 1958), a methylbetaine derivative of nicotinic acid, with mild hypoglycemic (Shani *et al.*, 1974; Bever and Zahnd, 1979; Marles and Farnworth, 1994) and antipellagra action (Covello, 1943; Bever and Zahnd, 1979) is the main N-compound of the seeds. Raw and dry fenugreek seeds contain about 0.15 percent of trigonelline and practically no nicotinic acid. If the seeds are sufficiently roasted about 2/3 of trigonelline is converted into nicotinic acid (Covello, 1943). A higher value of *c*. 0.38 percent for trigonelline and *c*. 0.003 percent for nicotinic acid content has also been reported (Kühn and Gerhard, 1943).

Callus cultures contain 3–4 times more trigonelline than the seeds of the plant and 12–13 times more than the roots and shoots. Even higher levels of this compound were produced by suspension cultures (Radwan and Kokate, 1980). Choline was also found in the seeds (Karrer, 1958).

Trigonella corniculata L.

Choline and betaine were isolated from seeds, while trigonelline was not found (Atal and Sood, 1964).

Trigonella polycerata L.

Aerial parts, roots, seeds and callus cultures were analyzed for trigonelline content, which was found to be highest in the seeds (0.25 percent), compared to those of the aerial parts (0.20 percent) and roots (0.13 percent) (Mehra *et al.*, 1996).

Anthocyanins

Although certain glycosidic patterns of the anthocyanins are common (e.g. 3-glucosides, 3,5-diglucosides), there are many more complex patterns with a variety of other sugars that are of more restricted occurrence. Such glycosidic patterns may show correlations with taxonomy. A rare type of glycoside, in which the 3-sugar is rhamnose instead of glucose, occurs in the *Trigonella* species. The presence in the petal pigments of anthocyanidin-3-rhamnoside-5-glucosides provides a character that differentiates *Trigonella* from two other genera belonging to the same tribe, Trifolieae, namely *Medicago* and *Trifolium*, whose petals contain anthocyanidin-3,5-diglucosides (Harborne and Turner, 1984).

Flavonoids

Trigonella species are rich in flavonoids (Harborne, 1971).

Trigonella foenum-graecum L.

Quercetin (quercetin-3-0-rhamnoside) has been reported (Gánju and Puri, 1959). The seeds of the plant were found to contain luteolin, quercetin (Varshney and Sharma, 1966), vitexin (8-C-β-D-glucosyl 5,7,4′-trihydroxyflavone), vitexin-7-0-glucoside (afroside), arabinoside of orientin or isoorientin (8-C-/6-C-β-D-glucosyl-arabinosyl-5,7,3′,4′-tetrahydroxyflavone) (Adamska and Lutomski, 1971). Wagner *et al.* (1973) confirmed the presence of vitexin and reported the isolation of isovitexin (saponaretin), isoorientin (6-C-glucosyl-luteolin), vicenin-1 (6-C-β-D-xylopyranosyl-8-C-β-D-glucopyranosyl-apigenin) in considerable quantities and vicenin-2 (6,8-C-β-D-diglucosylapigenin). Ecological factors may play a role in the varied occurrence of vicenin-1 and vicenin-2 in fenugreek seeds of different origin.

Vitexin-2″-0-p-coumarate was also isolated from fenugreek seeds (Sood *et al.*, 1976).

Vitexin, orientin (Huang and Liang, 2000), quercetin, naringenin, tricin and tricin-7-0-β-D-glucopyranoside (Shang *et al.*, 1998) were isolated from seeds originating from China. The last three flavonoids were isolated from fenugreek for the first time.

The presence of 4′,7-dihydroxyflavone, 3′,4′,7-trihydroxyflavone, formononetin (7-hydroxy,4′-methoxy-isoflavone), kaempferol-3-0-glycoside, kaempferol-3,7-diglycoside (the nature of glycosylation is uncommon), kaempferol-3,7-0-diglucoside, quercetin-3-0-glucoside (isoquercitrin) and quercetin-3,7-diglucoside has been reported (unspecified parts; Saleh *et al.*, 1982).

Luteolin, quercetin, vitexin, isovitexin and 7,4′-dimethoxyflavanone were isolated from an alcoholic extract of the whole plant (Parmar *et al.*, 1982); kaempferol and quercetin from a leaf extract (Sood, 1975). Isorhamnetin (3′-methoxy-quercetin) and kaempferol were found in hydrolysates from leaves (Daniel, 1989), while quercetin and kaempferol were detected in hydrolysates from flowers; these aglycones are the most common in the flowers of several *Trigonella* species (Jurzysta *et al.*, 1988). Investigation of the stems resulted in the isolation of the luteolin, quercetin and vitexin (Khurana *et al.*, 1982).

Recently, the following flavonol glycosides have been isolated from the fenugreek stems growing in China: kaempferol 3-0-β-D-glucosyl (1→2)-β-D-galactoside, kaempferol 3-0-β-D-glucosyl (1→2)-β-D-galactoside 7-0-β-D-glucoside, kaempferol 3-0-β-D-glucosyl (1→2)-(6″-0-acetyl)-β-D-galactoside 7-0-β-D-glucoside, quercetin 3-0-β-D-glucosyl (1→2)-β-D-galactoside 7-0-β-D-glucoside and kaempferol 3-0-β-D-glucosyl (1→2)-β-D-galactoside (Han *et al.*, 2001).

Luteolin, quercetin and vitexin-7-glucoside (afroside) were also isolated from 36 months old unorganized seedling callus tissue. The maximum flavonoid content was found in the fourth week of tissue growth (Uddin *et al.*, 1977).

An enhanced yield of luteolin, kaempferol, quercetin and vitexin was observed when the seeds were treated with low concentrations of chemical mutagens (Jain and Agrawal, 1990; Jain *et al.*, 1992).

The antibacterial activity shown by fenugreek seed extracts may be due to its flavonoid content (Bhatti *et al.*, 1996).

Other *Trigonella* species

Seshadri *et al.* (1972) reported the presence of two C-glycosides, identified as acacetin-6,8-di-C-glucoside and its monoacetate in the seeds of *T. corniculata* L. Bouillant *et al.* (1975) revised the structure and proposed that of 6-C-pentosyl, 8-C-hexosylacacetin. Vitexin was also isolated from the seeds (Seshadri *et al.*, 1973). Kaempferol, quercetin and myricetin were detected in hydrolysates from fresh flowers (Jurzysta *et al.*, 1988).

Hydrolysates from fresh flowers of *T. polycerata* L., *T. monspeliaca* L., *T. noëana* Boiss. and of *T. geminiflora* Boiss. contain kaempferol, quercetin, myricetin and laricytrin (3′-methoxy-myricetin), while those of *T. calliceras* Fisch. ex Bieb. and *T. cretica* (L.) Boiss. contain kaempferol, quercetin and myricetin and those of *T. coerulea* (L.) Ser. contain only the first two aglycones (Jurzysta *et al.*, 1988).

The aerial parts of *T. grandiflora* Bunge contain the C-glucosides orientin and vitexin and those of *T. tenuis* Fisch. ex Bieb. contain only vitexin (Bandyukova *et al.*, 1985).

The seeds of *T. occulta* Ser. were found to contain quercetin (Jain, 1976b).

Apigenin, luteolin, kaempferol and quercetin were also isolated from tissue cultures of *T. polycerata* L.; the flavonoid content was higher in the cultures than in the normal stage. Among the individual flavonoids, luteolin was at a maximum whereas kaempferol was at a minimum. Apigenin was absent in the root (Kamal and Yadav, 1991).

T. spicata Smith and Sm. (=*T. hamosa* Bess.) contains 4′,7-dihydroxyflavone, 3′,4′,7-trihydroxyflavone and their 7-β-O-glucopyranosides. The presence of kaempferol-3-robinobioside (biorobin) is not surprising, as this plant shows a morphological resemblance to *Melilotus* genus, which is also reported to contain 3-robinobiosides. Formononetin was also detected (unspecified parts; Saleh *et al.*, 1982).

T. coerulescens (Bieb.) Hal. contains kaempferol-3-glycoside and quercetin-3-O-glucoside (isoquercitrin), 4′,7-dihydroxyflavone, 3′,4′,7-trihydroxyflavone and the 7-O-glucoside of the latter flavone (unspecified parts; Saleh *et al.*, 1982).

Kaempferol, quercetin, 4′,7-dihydroxyflavone, 3′,4′,7-trihydroxyflavone and formononetin was stated to be present in the following species: *T. culindracea* Desv., *T. maritima* Del. ex Poir.; *T. anguina* Del., *T. monspeliaca* L., *T. laciniata* L. contain the same flavonoids, except quercetin (unspecified parts; Saleh *et al.*, 1982).

The 3,7-diglucosides of kaempferol and quercetin were found in *T. culindracea* Desv., while *T. anguina* Del. contain only the 3,7-diglycoside of kaempferol (unspecified parts; Saleh *et al.*, 1982).

The flavonoid content of eight Egyptian *Trigonella* species belonging to four different sections was investigated for chemotaxonomic purposes (Kawashty *et al.*, 1998).

Sect. Falcatulae Boiss.

T. maritima Poiret, *T. laciniata* L., *T. glabra* Thunb., *T. stellata* Forssk. were found to contain kaempferol 3-galactoglucoside, kaempferol 3,7-diglucoside, quercetin 3-galactoglucoside, 7,4′-dihydroxyflavone, 7,3′,4′-trihydroxyflavone, quercetin 7-diglucoside-3-*p*-coumaroylglucoside and the isoflavonoid formononetin.

Sect. Cylindracea Boiss.

In addition to the previously mentioned flavonoids *T. cylindracea* Boiss. was proved to contain quercetin 3,7-diglucoside and traces of kaempferol 7-glucoside and quercetin 7-glucoside.

Sect. Foenum-graecum L.

From *T. foenum-graecum* L. were isolated kaempferol 3-glucoside, kaempferol 7-glucoside, kaempferol 3-galactoglucoside, kaempferol 7-diglucoside-3-*p*-coumaroylglucoside, quercetin

3-glucoside, quercetin 7-glucoside, quercetin 3,7-diglucoside, quercetin 3-galactoglucoside, 7,4'-dihydroxyflavone, 7,3',4'-trihydroxyflavone and formononetin, while from *T. polyceratia* L. were isolated kaempferol 7-glucoside, kaempferol 3,7-diglucoside, quercetin 3-galactoglucoside, quercetin 3,7-diglucoside, formononetin and traces of kaempferol 3-glucoside, quercetin 3-glucoside and quercetin 7-glucoside.

Sect. Pectinatae Boiss.

From *T. arabica* Del. were isolated kaempferol 3-galactoglucoside, kaempferol 3,7-diglucoside, quercetin 3-galactoglucoside, quercetin 3,7-diglucoside, quercetin 7-diglucoside-3-*p*-coumaroylglucoside, 7,4'-dihydroxyflavone, 7,3',4'-trihydroxyflavone and formononetin.

Isoflavonoid phytoalexins

It is well known that many higher plants respond to microbial invasion by the *de novo* production of organic substances, phytoalexins. These compounds are absent from healthy plants and induced by the attacking micro-organisms. Different plant families accumulate chemically different types of compounds. Thus, the Leguminosae in general produce "induced isoflavonoids" (Ingham and Harborne, 1976). All the isoflavonoid phytoalexins thus far described accumulate as aglycones rather than as glycosides, the most regularly encountered compounds being phenolic pterocarpans and isoflavans. In contrast to pterocarpans and isoflavans, isoflavone and isoflavanone phytoalexins are limited, in terms of their distribution, within the Papilionoideae (Ingham, 1983).

The *Trigonella* species are divisible into three major groups on the basis of phytoalexin accumulation (Ingham, 1981). This phytoalexin approach to the study of systematic relationships within four related genera (*Medicago, Melilotus, Trifolium* and *Trigonella*) of the tribe Trifolieae enables the link of *Trigonella* to *Melilotus* on the one hand and to the *Medicago* on the other, while the third group (characterized by formation of maackiain) provides evidence for a connection to *Trifolium*.

The encountered phytoalexins are pterocarpan [medicarpin and maackiain] and isoflavan [vestitol and sativan] derivates (Figure 9.10). In *T. calliceras* Fisch. medicarpin was accompanied by a phytoalexin (designated TC-1), partially identified as a hydroxylated pterocarpan. Traces of three pterocarpan precursors, namely the isoflavone formononetin, the flavanone liquiritigenin and the chalcone isoliquiritigenin accompanied the above phytoalexins in a few species.

The grouping of *Trigonella* species based on their phytoalexin production (Ingham and Harborne, 1976; Ingham, 1981) is the following:

— *Group 1a* (medicarpin in quantity): *T. anguina* Del., *T. arabica* Del., *T. aristata* Vass., *T. balansae*, Boiss. ex Reut., *T. caelesyriaca* Boiss., *T. corniculata* L., *T. cretica* (L.) Boiss., *T. hamosa* L.,

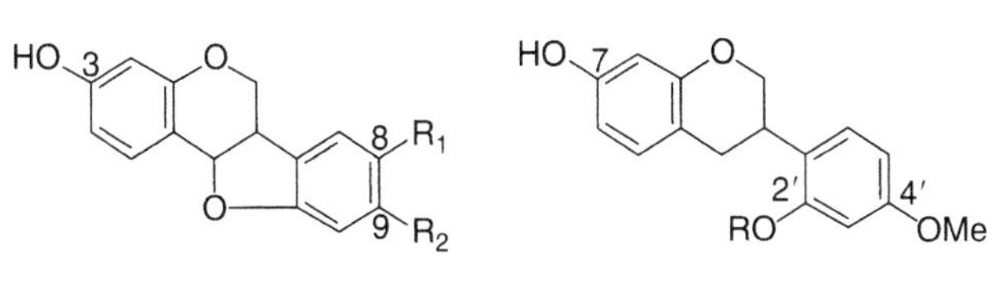

| Medicarpan $R_1 = H; R_2 = OMe$ | Vestitol $R = H$ |
| Maackiaian $R_1 = R_2 = O\text{-}CH_2-$ | Sativan $R = Me$ |

Figure 9.10 Chemical structures of commonly encountered Isoflavonoid Phytoalexins in *Trigonella* species.

T. pamirica Gross. in Kom., *T. rigida* Boiss. and Bal., *T. schlumbergeri* Buser (*sic*, possibly Boiss.), *T. spicata* Sibth and Sm., *T. spinosa* L., *T. stellata* Forssk., *T. suavissima* Lindl., *T. uncata* Boiss. and Noe [=*T. glabra* subsp. *uncata* (Boiss. and Noe) Lassen];

– *Group 1b* (medicarpin in traces): *T. lilacina* Boiss. and *T. monspeliaca* L.;

– *Group 1c* (medicarpin plus TC-1): *T. calliceras* Fisch.

– *Group 2a* (medicarpin + vestitol): *T. brachycarpa* Moris, *T. noëana* Boiss., *T. radiata* Boiss.;

– *Group 2b* (medicarpin + vestitol + sativan): *T. arcuata* C.A. Meyer, *T. cancellata* Dest., *T. fischeriana* Ser., *T. geminiflora* Boiss., *T. incisa* Benth., *T. monantha* C.A., Meyer, *T. orthoceras* Kar. and Kir., *T. platycarpos* L., *T. polycerata* L., *T. popovii* Kor., *T. ruthenica* L., *T. tenuis* Fisch.

– *Group 3a* (ratio 1 : 1 of medicarpin and maackiain): *T. berythaea* Boiss. ex Bl., *T. foenum-graecum* L., *T. gladiata* Stev.;

– *Group 3b* (ratio 10 : 1 of medicarpin and maackiain): *T. coerulea* (L.) Ser., *T. caerulescens* (Bieb.) H., *T. cylindracea* Desv., *T. kotschyi* Fenzl., *T. melilotus-caerulea*[2] (L.) Ascherson et Graebner, *T. procumbens* (Besser) Reichb., *T. sibthorpii* Boiss.(= *T. spruneriana* Boiss.).

It has been proved that in fenugreek seedlings medicarpin [(6aR, 11aR)-demethylhomoptero-carpin) is synthesized from 2′,7-dihydroxy-4′-methoxy-isoflavone via an overall *trans* addition of hydrogen to the double bond (Dewick and Ward, 1977).

Other phenolic compounds

Trigonella foenum-graecum L.

Scopoletin, chlorogenic, caffeic and *p*-coumaric acids were found in root, shoot and pod (Reppel and Wagenbreth, 1958); scopoletin and the lignan γ-schisandrin were found in leaves and stems (Wang *et al.*, 1997).

Hymecromone (4-methyl-7-acetoxycoumarin) was isolated from a whole plant extract for the first time (Bhardwaj *et al.*, 1977). The stems contain, besides hymecromone, (E)-3-(4-hydroxyphenyl)-2-propenoic acid (*p*-coumaric acid) and trigoforin (3,4,7-trimethyl-coumarin); the latter was isolated for the first time from this source (Khurana *et al.*, 1982).

Trigocoumarin, whose structure was first assigned as 3-(ethoxycarbonyl) methyl-4-methyl-5,8-dimethoxycoumarin was also isolated for the first time from a whole plant extract, together with hymecromone (Parmar *et al.*, 1982). The structure was further revised and the compound was assigned as 3-(ethoxycarbonyl)methyl-4-methyl-7,8-dimethoxycoumarin (Parmar *et al.*, 1984).

T. corniculata L.

Aesculetin and scopoletin (shoots; Reppel and Wagenbreth, 1958).

T. coerulea (L.) Ser.

Aesculetin, scopoletin and coumarin (shoots; Reppel and Wagenbreth, 1958). Due to its coumarin content, it is diuretic, digestive, antispasmodic and slightly hypnotic (Fournier 1948).

T. calliceras Fisch. ex Bieb.

Aesculetin, umbelliferon (shoots; Reppel and Wagenbreth, 1958).

T. cretica (L.) Boiss.

Aesculetin (shoots; Reppel and Wagenbreth, 1958).

Volatile constituents

From about fifty-one detected volatile constituents of *T. foenum-graecum* L. seeds, thirty-nine have been identified; among them *n*-alcanes, sesquiterpenes (the most important are elemenes and muurolenes) and some oxygenated components. The identified compounds are the following: *n*-hexenol, 2-heptanone, n-heptanal, aniline, phenol, heptanoic acid, 3-octen-2-one, 1,8-cineol, undecane, camphor, 5-methyl-δ-caprolactone, 1-dodecene, methylcyclohexylacetate, dihydrobenzofuran, dodecane, decanoic acid, thymol, 2-hexylfuran, tridecane, γ-nonalactone, eugenol, δ-elemene, 1-tetradecene, tetradecane, calarene, β-ionone, α-muurolene, dihydro-actinidiolide, ε-muurolene, β-elemene, β-selinene, γ-elemene, γ-muurolene, calamenene, pentadecane, dodecanoic acid, diphenylamine, 1-hexadecene, hexadecane. (Girardon *et al.*, 1985).

The contribution of *n*-alcanes, ranging from undecane to hexadecane to the aroma of fenugreek seeds was considered minimal. Although γ-nonalactone and 5-methyl-δ-caprolactone are present in small quantities, these compounds could be of great importance in the aroma of seeds because of their olfactory properties, but they do not possess *a priori* the characteristic and persistant odor of the seeds.

This odor is attributed to an oxygen heterocycle identified as 3-hydroxy-4,5-dimethyl-2(5H)-furanone, called HDMF, previously isolated from other sources such as the yellow wine of Jura, the melassa from sugar cane, etc. In all cases, it was proved to be the "key" component of their aroma. It is a polar, thermolabile compound, difficult to detect, when present in low concentrations (Girardon *et al.*, 1986).

Amino acids

The principal free amino acid of *T. foenum-graecum* L. was found to be (2S,3R,4R)-4-hydroxy-isoleucine [2-amino-4-hydroxy-3-methylpentanoic acid; (2S, 3R, 4R) form]. The (2R,3R,4S) isomer forms a minor component of fenugreek seed. Judging from the mild isolation procedure, it is unlikely to be an artifact (Hatanaka, 1992). The total amount of 4-hydroxyisoleucine present in the plant increases steadily during all phases of growth (Fowden *et al.*, 1973; Hardman and Abu-Al-Futuh, 1979).

Studies have shown that hydroxyisoleucine represents up to 80 percent of free amino acid in fenugreek dry seeds. The concentration does not decrease in the later stages of maturation of the seed, but it is absent from the seed reserve proteins (Sauvaire *et al.*, 1984). Fowden *et al.* (1973) estimated this amino acid level to be about 30–50 percent of the dry seeds' total free amino acid content.

The stereochemistry of the 4-hydroxyisoleucine from fenugreek has been reinvestigated and the absolute configuration was shown to be (2S, 3R, 4S) (Alcock *et al.*, 1989).

Although it has been reported in a preliminary test that the free (2S,3R,4S)-4-hydroxy-isoleucine shows no hypoglycemic activity (Hardman and Abu-Al-Futuh, 1976), it was proved that this amino acid possesses insulin-stimulating properties both *in vitro* and *in vivo* (Sauvaire and Ribes, 1993).

It is interesting that purified 4-hydroxyisoleucine alone gives the same aroma as 3-hydroxy-4,5-dimethyl-2(5H)-furanone (Hatanaka, 1992). This amino acid is probably the potential precursor of HDMF, through a oxidative desamination reaction (Girardon *et al.*, 1986). It is present in most species of *Trigonella* genus except *T. cretica* (L.) Boiss. Its formation is dependent on the presence of Fe^{2+}, 2-oxoglutarate, ascorbate, oxygen and of a 2-oxoacid dioxygenase (Haefelé *et al.*, 1997).

The nitrogen-rich non-protein amino acid, (S)-canavanine, although it seems to be unique to the Leguminosae subfamily Papilionoideae, does not occur in all species or genera (Tschiersch, 1959). It is particularly abundant in a free state in the tribe Trifolieae. It was found in *Trigonella arabica* Del., *T. coerulea* (L.) Ser. and *T. foenum-graecum* L., while it is absent from *T. berythaea* Boiss. ex Bl., *T. schlumbergeri* Boiss. and *T. stellata* Forssk. (Birdsong *et al.*, 1960; Bell *et al.*, 1978).

T. foenum-graecum L. seeds contain 1.5 g of canavanine per 16 g N (Van Etten *et al.*, 1961).

Fenugreek is one of the most important leafy vegetables consumed in India, whose seeds contain about 25–30 percent protein (Rao and Sharma, 1987; Sauvaire *et al.*, 1984; Duke, 1986). It is a promising crop giving more than 55 percent extractable protein N, at a rate of 0.3 kg/ha/day (Deshmukh *et al.*, 1974).

The amino acid composition of fenugreek seed protein in mg/g N was found to be (Hidvégi *et al.*, 1984): asp 672, thr 226, ser 276, glu 883, pro 292, gly 246, ala 212, cys 75, val 186, met 54, ileu 250, leu 361, tyr 167, phe 257, lys 345, his 159, try 93, arg 524, met+cys 129, tyr+phe 424.

This amino acid pattern is characterized by a relatively low quantity of sulfur-containing amino acids (129 mg/g N), but the amino acid pattern of the protein, unlike that of cereals, is particularly rich in lysine (345 mg/g N) (Hidvégi *et al.*, 1984). The protein quality of fenugreek seeds is approximately equal to that of soybeans (Sauvaire *et al.*, 1984). Contradictory data are reported for tryptophane content, that is, low levels (23 mg/g N by Sauvaire *et al.*, 1984; 93 mg/g N by Hidvégi *et al.*, 1984) in contrast to the higher level referred by Duke (1986).

The hypocholesterolemic effect shown by defatted fenugreek seeds could be based either on its amino acid pattern or to the considerable proportion of fibers (53.9 percent) and saponins (4.8 percent) (Valette *et al.*, 1984).

Inorganic elements

Since trace elements act as catalysts in biochemical reactions in living cells and are dietary essentials for animals and human beings, their levels in *T. foenum-graecum* L. were estimated (Sherif *et al.*, 1979).

The essential elements Ba, Br, Co, Cu, Fe, Mn and Zn were found in sufficient amounts in fenugreek seed, while, Se, also an essential element, was not found. From the rare earth elements La, Ce, Sm were also detected, while Eu, Tb, Yb were absent. Cs and Sb were present, but at the detection limits (Ila and Jagam, 1980). The plant was found rich in Mg (Kansal and Pahwa, 1979), Ca and Fe (Talwalkar and Patel, 1962). Whole grain contain (per 100 g edible portion) is: 220 mg Ca, 358 mg P, 24.2 mg Fe, while flour contains 213 mg Ca, 270 mg P and 32.4 mg Fe. Raw leaves contain 150 mg Ca and 48 mg P/100 g (Duke, 1986). The Sr content was found to be less than 100 γ/g dry wt. (Sarkar and Chauhan, 1963; Chauhan and Sarkar, 1964).

Vitamins

The amounts of various vitamins in *T. foenum-graecum* L. were estimated to be (in γ/g seed): thiamine 2.5, riboflavine 10.0, pyridoxine 11.0, cyanocobalamin 0.00025, niacin 2.5, Ca pantothenate 7.5 and biotin 0.0037 (Picci, 1959). Other reports refer to the following levels (per 100 g edible portion): 55 μg β-carotene equivalent, 0.32 mg thiamine, 0.30 mg riboflavin, 1.5 mg niacin. Flour contains (per 100 g): 0.06? mg thiamine, 0.05? mg riboflavin, and 1.5 mg niacin (Duke, 1986). The ascorbic acid content was found in fresh leaves to be *c.* 276 mg/100 g

(Sreeramulu *et al.*, 1983). Studies on the localization of vitamin C in the different parts of *T. foenum-graecum* L. showed that the leaves and the rapidly growing tissues contained approximately 80 percent of the total vitamin content, while the stems and the roots gave low values (Venkataramani, 1950).

Lipids

The quantitive analysis of lipid classes and the patterns of their constituent fatty acids in the leaves of *T. foenum-graecum* L. revealed the following composition: monogalactosyldiglycerides (11.3 percent of total lipids) and digalactosyldiglycerides (9.7 percent) (linolenic acic is predominant in both), sulfoquinovosyldiglycerides (3.1 percent) (characterized by a high content of palmitic and linolenic acids, the former being predominant), phosphatidylcholines (10.6 percent), phosphatidylethanolamines (5.1 percent), phosphatidylglycerols (3.7 percent) and other phospholipids (1.8 percent). In the various phospholipid classes linoleic acid predominates. Phosphatidylglycerols represent the only class that contains considerable proportions of *trans*-3-hexadecenoic acid (10.9 percent) (Radwan, 1978).

Total lipids extracted from fenugreek seeds amounted to 7.5 percent of the dry material. The total lipids consisted of 84.1 percent neutral lipids, 5.4 percent glycolipids and 10.5 percent phospholipids. Neutral lipids consisted mostly of triacylglycerols (86 percent), diacylglycerols (6.3 percent) and small quantities of monoacylglycerols, free fatty acids and sterols. Acylmonogalactosyldiacylglycerol and acylated stearylglycoside were the major glycolipids, while stearylglucoside, monogalactosylmonoacylglycerol and digalactosyldiacylglycerol were present in small amounts. The phospholipids consisted of phosphatidylcholine and phosphatidylethanolamine, as major phospholipids and phosphatidylserine, lysophosphatidylcholine, phosphatidylinositol, phosphatidylglycerol and phosphotidic acid as minor phospholipids (Hemavathy and Prabhakar, 1989).

Aliphatic natural products and carbohydrates

T. corniculata L.

The seeds of the plant are reported to contain triacontane (Atal and Sood, 1964), ethyl-α-D-galactopyranoside (Varshney *et al.*, 1974), ethyl-β-D-galactopyranoside (Seshadri *et al.*, 1973), while the stems and the leaves are reported to contain D-pinitol (3-O-methyl-D-inositol) (Plouvier, 1955).

T. coerulea (L.) Ser.

The most important flavor components of this herb, used as flavoring in a special swiss cheese (Schabzieger), were found to be some α-keto acids: pyruvic acid (α-ketopropionic acid), (α-ketoglutaric acid (2-oxopentanedioic acid), α-ketoisovaleric acid and α-ketoisocaproic acid (unspecified parts; Ney, 1986).

References

Adamska, Von M. and Lutomski J. (1971) C-Flavonoidglykoside in den Samen von *Trigonella foenum-graecum* L. *Planta Med.*, **20**, 224–9.

Alcock, N.W., Crout, D.H.G., Gregorio, M.V.M., Lee, E., Pike, G. and Samuel, C.J. (1989) Stereochemisty of the 4-hydroxyisoleucine from *Trigonella foenum-graecum. Phytochemistry,* **28**(7), 1835–41.

André, J. (1956) *Lexique des termes de botanique en latin. Etudes et commentaires,* Librairie C. Klincksieck, Paris, **XXIII**, p. 135.

Atal, C.K. and Sood, S.P. (1964) A phytochemical investigation of *Trigonella corniculata* Linn. *J. Pharm. Pharmacol.,* **16**, 627–9.

Balbaa, S.I., Wahab, S.M., Abd El, Selim, M. and Abo El Fotouh (1977) Study of the sapogenin content of the different organs of *Melilotus siculus* (Turra) growing in Egypt. *Egypt. J. Pharm. Sci.,* **18**(3), 293–304. CA 94: 27378g.

Bandyukova, V.A., Khalmatov, Kh. Kn. and Yunusova, K.K. (1985) Flavonoids of *Trigonella grandiflora* and *T. tenuis. Khim. Prir. Soedin.,* 4, 562–3.CA 104: 65961z.

Bedour, M.S., El-Munajjed, D., Fayez, M.B.E. and Girgis, A.N. (1964) Steroid Sapogenins VII. Identification and origin of 25D-Spirosta-3,5-diene among the fenugreek sapogenins. *J. Pharm. Sci.,* **53**(10), 1276–8.

Bell, E.A., Lackey, J.A. and Polhill, R.M. (1978) Systematic Significance of Canavanine in the Papilioideae (Faboideae). *Biochem. Syst. Ecol.,* 6, 201–12.

Bever, B.O. and Zahnd, G.R. (1979) Plants with oral hypoglycaemic action. *Quart. J. Crude Drug Res.,* 17(3–4), 139–96.

Bhardwaj, D.K., Murari, R., Seshadri, T.R. and Singh, R. (1977) Isolation of 7-acetoxy-4-methylcoumarin from *Trigonella foenum-graecum. Indian J. Chem., Sect. B,* 15(1), 94–5. CA 87: 19045m.

Bhatnagar, S.C., Misra, G., Nigam, S.K., Mitra, C.R. and Kapool, L.D. (1975) Diosgenin from *Balanites roxburghii* and *Trigonella foenum-graecum. Quart. J. Crude Drug Res.,* 13, 122–6.

Bhatti, M.A., Khan, M.T.J., Ahmed, B., Jamshaid, M. and Ahmad, W. (1996) Antibacterial activity of *Trigonella foenum-graecum* seeds. *Fitoterapia,* 67(4), 372–4.

Birdsong, B.A., Alston, R. and Turner, B.L. (1960) Distribution of canavanine in the family Leguminosae as related to phyletic groupings. *Can. J. Bot.,* 38, 499–505.

Blunden, G., and Hardman, R. (1963) *Dioscorea belizensis* Lundell as a source of diosgenin. *J. Pharm. Pharmacol.,* 15, 273–80.

Blunden, G. Hardman, R. and Wensley, W.R. (1965) Effects of enzymes on *Yucca glauca* Nutt. and other steroid-yielding monocotyledons. *J. Pharm. Pharmacol.,* 17, 274–80.

Bogacheva, N.G., Kiselev, V.P. and Kogan, L.M. (1976a) Isolation of 3,26-biglycoside of yamogenin from *Trigonella foenum-graecum. Khim.-Prir. Soedin.,* 2, 268–9. CA 85: 106634e.

Bogacheva, N.G., Ulezlo, I.V. and Kogan, L.M. (1976b) Steroid genins of *Trigonella foenum-graecum* seeds. *Khim.-Farm. Zh.,* 10(3), 70–2. CA 85: 68172t.

Bogacheva, N.G., Gorokhova, M.M., Kudryavtseva, V.N., Kiselev, V.P. and Kogan, L.M. (1976c) Steroid genins of *Trigonella coerulea* seeds. *Khim.-Farm. Zh.,* 10(8), 78–80. CA 86:52662n.

Bogacheva, N.G., Sheichenko, V.I. and Kogan, L.M. (1977a) Stucture of yamogenin tetroside from *Trigonella foenum-graecum* seeds. *Khim.-Farm. Zh.,* 11(7), 65–9. CA 87:180685e.

Bogacheva, N.G., Gorokhova, M.M. and Kogan, L.M. (1977b) 22-Methyl ether of protodioscin from *Trigonella coerulea* seeds. *Khim. Prir. Soedin.,* 3, 421. CA 87: 148665n.

Bohannon, M.B., Hagemann, J.W., Earle, F.R. and Barclay, A.S. (1974) Screening seed of *Trigonella* and three related genera for diosgenin. *Phytochemistry,* 13, 1513–4.

Bouillant, M.L., Favre-Bonvin, J. and Chopin, J. (1975) Structural determination of C-glycosylflavones by mass spectroscopy of their permethyl ethers. *Phytochemistry,* 14, 2267–74.

Brain, K.R. and Lockwood, G.B. (1976) Hormonal control of steroid levels in tissue cultures from *Trigonella foenum-graecum. Phytochemistry,* 15, 1651–4.

Brain, K.R., Fazli, F.R.Y., Hardman, R. and Wood, A.B. (1968) The rapid quantitative determination of C_{25} epimeric steroidal sapogenins in plants. *Phytochemistry,* 7, 1815–23.

Brenac, P. and Sauvaire, Y. (1996) Chemotaxonomic value of sterols and steroidal sapogenins in the genus *Trigonella. Biochem. Syst. Ecol.,* 24(2), 157–64.

Carnoy, A. (1959) *Dictionnaire étymologique des noms grecs de plantes.* Bibliothèque du Muséon, Louvain, 46, pp. 259.

Chauhan, U.P.S. and Sarkar, R.B.C. (1964) A microchemical method for the determination of strontium in some plant material. *Indian J. Chem.,* 2(5), 175–8. CA 61: 6026b.

Che, C.-T. (1991) Source of potential antiviral agents. In H. Wagner and N.R. Farnsworth (eds), *Economic and Medicinal Plant Research*, Academic Press, London, 5, pp. 215.

Cornish, M.A., Hardman, R. and Sadler, R.M. (1983) Hybridisation for genetic improvement in the yield of diosgenin from fenugreek seed. *Planta Med.*, 48, 149–52.

Covello, M. (1943) Trigonellin and nicotinic acid in *Trigonella foenum-graecum* and their relation to antipellagra activity. *Boll. Soc. Ital. Biol. Sper.*, 18, 159–61. CA 41: 797d.

Daniel, M. (1989) Polyphenols of some Indian vegetables. *Curr. Sci.*, 58(23), 1332–4.

Dawidar, A.M. and Fayez, M.B.E. (1972) Thin-layer chromatographic detection and estimation of steroid sapogenins. *Z. Anal. Chem.*, 259, 283–5.

Dawidar, A.M., Saleh, A.A. and Elmotei, S.L. (1973) Steroid sapogenin constituents of fenugreek seeds. *Planta Med.*, 24(4), 367–70.

Deshmukh, M.G., Gore, S.B., Mungikar, A.M. and Joshi, R.N. (1974) The yields of year protein from various short-duration crops. *J. Sci. Fd Agric.*, 25, 717–24.

Dewick, P.M. and Ward, D. (1977) Stereochemistry of isoflavone reduction during pterocarpan biosynthesis: on investigation using Deuterium Nuclear Magnetic Resonance Spectroscopy. *J. Chem. Soc. Chem. Comm.*, 338–9.

Duke, J.A. (1986) *Handbook of Medicinal Herbs*, CRC, Florida, p. 490.

Duke, J.A. (1992) *Handbook of Biologically Active Phytochemicals and Their Activities*, CRC, Florida, p. 63.

Elujoba, A.A. and Hardman, R. (1985a) Incubation conditions for fenugreek whole seed. *Planta Med.*, 51(2), 113–5.

Elujoba, A.A. and Hardman, R. (1985b) Fermentation of powdered fenugreek seeds for increased sapogenin yields. *Fitoterapia*, 66(6), 368–70.

Fazli, F.R.Y. (1967) Studies in steroid-yielding of the genus *Trigonella*. PhD Thesis, University of Nottingham.

Fazli, F.R.Y. and Hardman, R. (1968) The spice, fenugreek, (*Trigonella foenum-graecum* L.): its commercial varieties of seed as a source of diosgenin. *Tro. Sci.*, 10, 66–78.

Fazli, F.R.Y. and Hardman, R. (1971) Isolation and characterization of steroids and other constituents from *Trigonella foenum-graecum* L. *Phytochemistry*, 10, 2497–503.

Fournier, P. (1948) Trigonelle in *Le livre des Plantes médicinales et vénéneuses de France*, P. Lechevalier, III, pp. 494–501.

Fowden, L., Pratt, H. and Smith, A. (1973) 4-Hydroxyisoleucine from seed of *Trigonella foenum-graecum*. *Phytochemistry*, 12, 1707–11.

Gánju, K. and Puri, B. (1959) Bioflavonoids from Indian vegetables and fruits. *Indian J. Med. Res.*, 47, 563–570. CA 54: 2627b.

Ghosal, S., Srivastava, R.S., Chatterjee, D.C. and Dutta, S.K. (1974) Fenugreekine, a new steroidal sapogenin-peptide ester of *Trigonella foenum-graecum*. *Phytochemistry*, 13, 2247–51.

Girardon, P., Bessiere, J.M., Baccou, J.C. and Sauvaire. Y. (1985) Volatile constituents of fenugreek seeds. *Planta Med.*, 51,533–4.

Girardon, P., Sauvaire, Y., Baccou, J.C. and Bessiere, J.M. (1986) Identification of 3-hydroxy-4,5-dimethyl-2(5H)-furanone in aroma of fenugreek seeds (*Trigonella foenum-graecum* L.). *Lebensm.-Wiss. Technol.*, 19(1), 44–6.

Grangrade, H. and Kaushal, R. (1979) Fenugrin-B, a saponin from *Trigonella foenum-graecum* Linn. *Indian Drugs*, 16(7), 149. CA 91: 52711f.

Gupta, R.K., Jain, D.C. and Thakur, R.S. (1984) Furostanol glycosides from *Trigonella foenum-graecum* seeds. *Phytochemistry*, 23(11), 2605–7.

Gupta, R.K., Jain, D.C. and Thakur, R.S. (1985a) Furostanol glycosides from *Trigonella foenum-graecum* seeds. *Phytochemistry*, 24(10), 2399–401.

Gupta, R.K., Jain, D.C. and Thakur, R.S. (1985b) Plant saponins. IX. Trigofoenoside E-1, a new furostanol saponin from *Trigonella foenum-graecum*. *Indian J. Chem.. Sect. B*, 24(12), 1215–7.

Gupta, R.K., Jain, D.C. and Thakur, R.S. (1986a) Two furostanol glycosides from *Trigonella foenum-graecum*. *Phytochemistry*, 25(9), 2205–7.

Gupta, R.K., Jain, D.C. and Thakur, R.S. (1986b) Minor steroidal sapogenins from fenugreek seeds, *Trigonella foenum-graecum*. *J. Nat. Prod.*, 48, 1153.

Haefelé, C., Bonfils, C., Sauvaire, Y. (1997) Characterization of a dioxygenase from *Trigonella foenum-graecum* involved in 4-hydroxyisoleucine biosynthesis. *Phytochemistry*, 44(4), 563–6.

Han, Y., Nishibe, S., Noguchi, Y. and Jin, Z. (2001) Flavonol glycosides from the stems of *Trigonella foenum-graecum*. *Phytochemistry*, 58, 577–80.

Harborne, J.B (1971) Distribution of flavonoids in the Leguminosae. In J.B. Harborne, D. Boulter and B.L. Turner (eds), *Chemotaxonomy of the Leguminosae*, Academic Press, London and New York. pp. 31–71.

Harborne, J.B. and Turner, B.L (1984) Plant pigment. In *Plant Chemosystematics*, Academic Press, London, pp. 139–40.

Hardman, R. and Abu-Al-Futuh, I. (1976) The occurrence of 4-hydroxyisoleucine in steroidal sapogenin-yielding plants. *Phytochemistry*, 15, 325.

Hardman, R. and Abu-Al-Futuh, I. (1979) The detection of isomers of 4-hydroxy-isoleucine by the Jeol Amino Acid Analyser and by TLC. *Planta Med.*, 36, 79–84.

Hardman, R. and Brain, K.R. (1971) The effect of post harvest application of growth regulators on the yield of steroidal sapogenin from plant material. *Phytochemistry*, 10, 519–23.

Hardman, R. and Brain, K.R. (1972) Variations in the yield of total and individual 25-α and 25β-sapogenins on storage of whole seed of *Trigonella foenum-graecum*. *Planta Med.*, 21(4), 426–30.

Hardman, R. and Fazli, F.R.Y. (1972a) Methods of screening the genus *Trigonella* for steroidal sapogenins in genus *Trigonella*. *Planta Med.*, 21(2), 131–8.

Hardman, R. and Fazli, F.R.Y. (1972b) Studies in the steroidal saponin yield from *Trigonella foenum-graecum* seed. *Planta Med.*, 21(3), 322–8.

Hardman, R. and Fazli, F.R.Y. (1972c) Labelled steroidal sapogenins and hydrocarbons from *Trigonella foenum-graecum* by acetate, melovanate and cholesterol feeds to seeds. *Planta Med.*, 21(2), 188–95.

Hardman, R. and Jefferies, T.M. (1971) The determination of diosgenin and yamogenin in fenugreek seed by combined column chromatography and infrared spectrometry. *J. Pharm. Pharmac.*, 23(Suppl.), 231S–232S.

Hardman, R. and Jefferies, T.M. (1972) A combined column-chromatographic and infrared spectrophoto-metric determination of diosgenin and yamogenin in fenugreek seed. *Analyst*, 97, 437–41.

Hardman, R., Kosugi, J. and Parfitt, R.T. (1980) Isolation and characterization of a furostanol glycoside from fenugreek. *Phytochemistry*, 19, 698–700.

Hardman, R. and Sofowora, E.A. (1971) Effect of enzymes on the yield of steroidal sapogenin from the epicarp and mesocarp of *Balanites aegyptiaca* fruit. *Planta Med.*, 20(2), 124–9.

Hardman, R. and Stevens, R.G. (1978) The influence of NAA and 2,4-D on the steroidal fractions of *Trigonella foenum-graecum* static cultures. *Planta Med.*, 34, 414–19.

Hardman, R. and Wood, C.N. (1971a) The ripe fruits of *Balanites orbicularis* as a new source of diosgenin and yamogenin. *Phytochemistry*, 10, 887–9.

Hardman, R. and Wood, C.N. (1971b) The effect of ripening and aqueous incubation on the yield of diosgenin and yamogenin from the fruits of *Balanites pedicellaris*. *Planta Med.*, 20(4), 350–6.

Hatanaka, S.-I. (1992) Amino acids from mushrooms. In W. Herz, G.W. Kirby, R.E. Moore, W. Steglich and Ch. Tamm (eds), *Progress in the Chemistry of Organic Natural Products*, Springer-Verlag, Wien, New York, pp. 14–16.

Hemavathy, J. and Prabhakar, J.V. (1989) Lipid composition of fenugreek, *Trigonella foenum-graecum* seeds. *Food Chem.*, 31(1), 1–8.

Hidvégi, M., El-Kady, A., Lásztity, R., Békés, F., Simon-Sarkadi, L. (1984) Contributions to the nutritional characterization of fenugreek (*Trigonella foenum-graecum* L. 1753) *Acta Alimentaria*, 13, 315–24.

Hooker, J.D. and Jackson, B.D. (1960) *Index Kewensis. An Enumeration of the Genera and Species of Flowering Plants*. Clarendon Press, Oxford, 2, pp. 1116–8.

Huang, W.-Z. and Liang, X. (2000) Determination of two flavone glycosides in the seeds of *Trigonella foenum-graecum* L. *Zhiwu Ziyan Yu Huanjing Xuebao*, 9(4), 53–4. CA 134: 277883.

Ila, P. and Jagam, P. (1980) Multielement analysis of food spices by instrumental neutron activation analy-sis. *J. Radioanal. Chem.*, 57(1), 205–10.

Ingham, J.L. (1981) Phytoalexin induction and its chemosystematic significance. *Biochem. Syst. Ecol.*, 9(4), 275–81.

Ingham, J.L. (1983) Naturally occurring isoflavonoids. In W. Herz, H. Grisebach and G.W. Kirby (eds), *Progress in the Chemistry of the Organic Natural Products*, Springer-Verlag, Wien, New York, pp. 6–14.

Ingham, J.L. and Harborne, J.B. (1976) Phytoalexin induction as a new dynamic approach to the study of systematic relationships among higher plants. *Nature*, 260, 241–3.

Jain, S.C. (1976a) Steroidal sapogenins from *Trigonella occulta. Lloydia*, 39, 244–5.

Jain, S.C. (1976b) Phytochemical study of *Trigonella occulta* Delile seeds. *Indian J. Pharm.*, 38(1), 25–6. CA 85: 74891t.

Jain, S.C. and Agrawal, M. (1987) Effect of chemical mutagens on steroidal sapogenins in *Trigonella* species. *Phytochemistry*, 26(8), 2203–5.

Jain, S.C. and Agrawal, M. (1990) Effect of sodium azide on pharmaceutically active flavonoids in *Trigonella* species. *Indian J. Pharm. Sci.*, 52(1), 17–19.

Jain, S.C. and Agrawal, M. (1994) Effect of mutagens on steroidal sapogenins in *Trigonella foenum-graecum* tissue cultures. *Fitoterapia*, 65(4), 367–70.

Jain, S.C., Agrawal, M. and Vijayvergia, R. (1992) Regulation of pharmaceutically active flavonoids in *Trigonella foenum-graecum* by alkylating agents. *Fitoterapia*, 63(6), 539–41.

Jain, S.C., Rosenberg, H. and Stohs, S.J. (1977) Steroidal constituents of *Trigonella occulta* tissue cultures. *Planta Med.*, 31(2), 109–11.

Jefferies, T.M. and Hardman, R. (1976) An improved column-chromatographic quantitative isolation of diosgenin and yamogenin from plant crude extracts prior to their determination by infrared spectrophotometry. *Analyst*, 101, 122–4.

Jurzysta, M., Burda, S., Oleszek, W. and Ploszynski, M. (1988) The chemotaxonomic significance of laricytrin and medicagenic acid in the tribe Trigonelleae. *Can. J. Bot.*, 66, 363–7.

Kamal, R. and Yadav, R. (1991) Flavonoids from *Trigonella polycerata in vitro* and *in vivo. J. Phyt. Res.*, 4(2), 161–5.

Kansal, V.K. and Pahwa, A. (1979) Utilization of magnesium from leafy vegetables and cereals: effect of incorporation skim milk powder in the diets. *J. Nutr. Diet.*, 16(12), 453–9. CA 92: 16294r.

Karrer, W. (1958) *Konstitution und Vorkommen der organischen Planzenstoffe*. Birkhäuser Verlag, Basel und Stuttgart. p. 997, 1009.

Kawashty, S.A., Abdalla, M.F., Gamal El Din, E.M. and Saleh, N.A.M. (1998). The chemosystematics of Egyptian *Trigonella* species. *Biochem. System. Ecol.*, 26, 851–6.

Khanna, P. and Jain, S.C. (1973) Diosgenin, gitogenin and tigogenin from *Trigonella foenum-graecum* tissue cultures. *Lloydia*, 36, 96–8.

Khurana, S.K., Krishnamoorthy, V., Parmar, V.S., Sanduja, R. and Chawla, H. L. (1982) 3,4,7-trimethyl-coumarin from *Trigonella foenum-graecum* stems. *Phytochemistry*, 21, 2145–6.

Knight, J.C. (1977) Analysis of fenugreek by gas–liquid chromatography. *J. Chrom.*, 133, 222–5.

Kogan, L.M. and Bogacheva, N.G. (1978) Novel glycoside of furost-5-ene-3β,22,26-triol from *Trigonella foenum-graecum* and *Trigonella coerulea. Khim.-Prir. Soedin.*, 5, 39. CA 93: 182786m.

Kühn, A. and Gerhard, H. (1943) The trigonellin and nicotinic acid contents of semen foenugraeci. *Arch. Pharm.*, 281, 378–9. CA 39: 50402.

Lockwood, G.B. and Brain, K.R. (1976) Influence of hormonal supplementation on steroid levels during callus induction from seeds of *Trigonella foenum-graecum. Phytochemistry*, 15, 1655–60.

Manha, S.K., Raisinghani, G. and Jain, S.C. (1994) Diosgenin production in induced mutants of *Trigonella corniculata. Fitoterapia*, 65(6), 515–6.

Marker, R.E., Wagner, R.B., Ulshafer, F.R., Goldsmith, D.P.J. and Ruof, C.H. (1943) Sterols CLIV. Sapogenins LXVI. The sapogenin of *Trigonella foenum-graecum. J.A.C.S.*, 65, 1247.

Marker, R.E., Wagner, R.B., Ulshafer, Wittbecker, E.L., Goldsmith, D.P.J. and Ruof, C.H. (1947) New sources for sapogenins. *J.A.C.S.*, 69, 2242.

Marles, R.J. and Farnsworth, N.R. (1994). Plants as sources of antidiabetic agents. In H. Wagner and N.R. Farnsworth (eds), *Economic and Medicinal Plant Research*, Academic Press London. 6, pp. 164–5.

Mehra, P., Yadav, R. and Kamal, R. (1996) Influence of nicotinic acid on production of trigonelline from *Trigonella polycerata* tissue culture. *Indian J. Exp. Biol.*, 34(11), 1147–9.

Murakami, T., Kishi, A., Matsuda, H. and Yoshikawa, M. (2000) Medicinal foodstuffs. XVII. Fenugreek seed. (3): Structures of new furostanol-type steroid saponins, trigoneosides Xa, Xb, XIb, XIIa, XIIb and XIIIa from seeds of Egyptian *Trigonella foenum-graecum* L. *Chem. Pharm. Bull.*, 48(7), 994–1000.

Ney, K.H. (1986) Investigation of the flavor of ziegerklee (*Coerulea mellilotus*) i.e. the key components of Schabzieger (special Swiss cheese with herbs). *Gordian*, 86 (1–2), 9–10. CA 105: 5347q.

Parmar, V.S., Jha, H.N., Sanduja, S.K. and Sanduja, R. (1982) Trigocoumarin – a new coumarin from *Trigonella foenum-graecum*. *Z. Naturforsch.*, 37b, 521–3.

Parmar, V.S., Singh, S. and Rathore, J.S. (1984) A structure revision of trigocoumarin. *J. Chem. Res. Synop.*, 11, 378. CA 103: 6079z.

Picci, G. (1959) Microbiological determinations of some vitamins and amino acids liberated during germination of the seeds. *Ann. fac. agrar. univ. Pisa*, 20, 51–60. CA 54: 17564b.

Plouvier, V. (1955) Pinitol in legumes. Quercitol in Pterocarpus lucens. *Compt. rend.*, 241, 1838–40. CA 44: 6485i.

Puri, H.S., Jefferies, T.M. and Hardman, R. (1976) Diosgenin and Yamogenin levels in some Indian plant samples. *Planta Med.*, 30, 118–21.

Radwan, S.S. (1978) Coupling of two-dimensional thin-layer chromatography with gas chromatography for the quantitive analysis of lipids classes and their constituent fatty acids. *J. Chrom. Sci.*, 16, 538–41.

Radwan, S.S. and Kokate, C.K. (1980) Production of higher levels of trigonellin by cell cultures of *Trigonella foenum-graecum* than by the differentiated plant. *Planta*, 147, 340–4.

Rao, P.U. and Sharma, R.D. (1987) An evaluation of protein quality of fenugreek seeds (*Trigonella foenum-graecum*) and their supplementary effects. *Food Chem.*, 24(1), 1–9. CA 107: 76312b.

Reppel, L. and Wagenbreth, D. (1958) Untersuchungen über den Gehalt an Cumarinen und diesen verwandten Säuren in Pfropfungen zwischen *Melilotus albus* Med. und *Trigonella foenum-graecum*. *Flora*, 146, 212–27.

Saleh, N.A.M., Boulos, L., El-Negoumy, S.I. and Abdalla, M.F. (1982) A comparative study of the flavonoids of *Medicago radiata* with other *Medicago* and related *Trigonella* species. *Biochem. Syst. Ecol.*, 10(1), 33–6.

Sarkar, B.C. and Chauhan, U.P.S. (1963) Strontium in some Indian vegetables. *Curr. Sci.*, 32(9), 418–9. CA 60: 1038g.

Sauvaire, Y. and Baccou, J.C. (1978) L' obtention de la Diosgénine, (25R)-Spirost-5-ène-3β-ol; Problèmes de l' hydrolyse acide des saponines. *Lloydia*, 41(3), 247–56.

Sauvaire, Y., Girardon, P., Baccou, J.C. and Ristérucci (1984) Changes in the growth, proteins and free amino acids of developing seed and pod of fenugreek. *Phytochemistry*, 23(3), 479–86.

Sauvaire, Y. and Ribes, G. (1993) French Patent No. 9210644, European Patent No. 93401353, US Patent No. 81113951.

Seshadri, T.R., Sood, A.R. and Varshney, I.P. (1972) Glycoflavones from the seeds of *Trigonella corniculata*. Isolation of 6,8-di-C-β-D-glucopyranosylacacetin and its monoacetate. *Indian J. Chem.*, 10(1), 26–8. CA 77: 2767u.

Seshadri, T.R., Varshney, I.P. and Sood, A.R. (1973) Glycosides from *Trigonella corniculata* and *Trigonella foenum-graecum* Linn. seeds. *Curr. Sci.*, 42(12), 412–2. CA 79: 102757a.

Shang, M., Cai, S., Han, J., Li, J., Zhao, Y., Zheng, J., Namba, T., Kadota, S., Tezuka, Y. and Fan, W. (1998) Studies on flavonoids from fenugreek. *Zhongguo Zhongyao Zazhi*, 23(10), 614–16. CA 130: 220364.

Shang, M., Tezuka, Y., Cai, S., Li, J., Kadota, S., Fan, W. and Namba, T. (1998) Studies on triterpenoids from common fenugreek. *Zhongcaoyao*, 29(10), 655–7. CA 130: 150917.

Shani, J., Goldschmied, A., Joseph, B., Ahronson, Z. and Sulman, F.G. (1974) Hypoglycaemic effect of *Trigonella foenum-graecum* and *Lupinus termis* (Leguminosae) seeds and their major alkaloids in alloxan-diabetic and normal rats. *Arch. Int. Pharmacodyn. Ther.*, 210(10), 27–37. CA 83: 90765u.

Sharma, R.D. (1986) An evaluation of hypocholesterolemic factor of fenugreek seeds (*Trigonella foenum-graecum*) in rats. *Nutr. Rep. Int.*, 33(4), 669–677. CA 104: 206054v.

Sherif, M.K., Awadallah, R.M. and Mohaned, A.E. (1979) Determination of trace elements of Egyptian crops by neutron activation analysis. II. Trace elements in Umbelliferae and Leguminosae families. *J. Radioanal. Chem.*, 53(1–2), 145–53.

Soliman, G. and Mustafa, Z. (1943) The saponin of fenugreek seeds. *Nature*, 151, 195–6.

Soliman, G. and Mustafa, Z. (1949) The saponins of fenugreek seeds. *Rept. Pharm. Soc. Egypt.*, 31, 119.

Sood, A.R. (1975) Chemical components from the leaves of *Trigonella foenum-graecum. Indian J. Pharm.*, 37(4), 100–1. CA 84: 40731e.

Sood, A.R., Boutard, B., Chadenson, M., Chopin, J. and Lebreton, P. (1976) A new flavone C-glycoside from *Trigonella foenum-graecum. Phytochemistry*, 15, 351–2.

Sreeramulu, N., Banyikwa, F.F. and Srivastava, V. (1983) Loss of ascorbic acid due to wilting in some green leafy vegetables. *J. Plant Foods*, 5(4), 215–9. CA 101: 228764s.

Talwalkar, R.T. and Patel, S.M. (1962) Nutritive value of some leaf proteins. I. Amino acid composition of *Trigonella foenum-graecum* and *Hibiscus cannabinus. Ann. Biochem. Exptl. Med.*, 22, 289–94. CA 58: 7300e.

Tschesche, S., Seidel, L., Sharma, S.C. and Wulff, G.(1972) Über Lanatigosid und Lanagitosid, zwei bisdesmolidische 22-Hydroxy-furostanol-Glycoside aus den Blättern von *Digitalis lanata* Ehrh. *Chem. Ber.*, 105, 3397–406.

Tschiersch, B. (1959) Über Canavanin. *Flora*, 147(3), 405–16.

Uddin, A., Sharma, G.L. and Khanna, P. (1977) Flavonoids from *in vitro* seedling callus culture of *Trigonella foenum-graecum* Linn. *Indian J. Pharm.*, 39(6), 142–3. CA 88: 47528k.

Valette, G., Sauvaire, Y., Baccou, J.C. and Ribes, G. (1984) Hypocholesterolemic effect of fenugreek seeds in dogs. *Atherosclerosis*, 50(1), 105–11.

Van Etten, C.H., Miller, R.W., Wolff, I.A. and Jones, Q. (1961) Amino acid composition of twenty-seven selected seed meals. *J. Agr. Fd Chem.*, 9(1), 79–82.

Varshney, I.P. and Beg, M.F.A. (1978) Study of saponins from the seeds of *Trigonella foenum-graecum* Linn. *Indian J. Chem.. Sect. B* 16(12), 1134–6. CA 91: 87294z.

Varshney, I.P. and Jain, D.C. (1979) Study of glycosides from *Trigonella foenum-graecum* Linn. leaves. *Natl. Acad. Sci. Lett.*, 2(9), 331–2. CA 93: 66061x.

Varshney, I.P., Jain, D.C. and Srivastava, H.C. (1984) Saponins from *Trigonella foenum-graecum* leaves. *J. Nat. Prod.*, 47(1), 44–6.

Varshney, I.P., Jain, D.C., Srivastava, H.C., Singh, P.P. (1977) Study of saponins from *Trigonella foenum-graecum* Linn. leaves. *J. Indian Chem. Soc.*, 54(12), 1135–6. CA 89: 126108x.

Varshney, I.P. and Sharma, S.C. (1966) Saponins and sapogenins. XXXII. *Trigonella foenum-graecum* seeds. *J. Indian Chem. Soc.*, 43(8), 564–7. CA 65: 18991.

Varshney, I.P. and Sood, A.R. (1969) Sapogenins from *Trigonella corniculata* Linn. *J. Indian Chem. Soc.*, 46(5), 391–2. CA 71: 42195p.

Varshney, I.P. and Sood, A.R. (1971) Sapogenins from *Trigonella foenum-graecum* stems and leaves and *T. corniculata* leaves and flowers. *Indian J. Appl. Chem.*, 34(5), 208–10. CA 77: 85529s.

Varshney, I.P., Sood, A.R., Srivastava, H.C. and Harshe, S.N. (1974) Isolation of ethyl galactoside from *T. corniculata* seeds. *Planta Med.*, 26, 26–32.

Venkataramani, K.S. (1950) The factors governing the vitamin C content of *Trigonella foenum-graecum. Proc. Indian Acad. Sci.*, 32B, 112–125. CA 45: 3465b.

Wagner, H., Iyengar, M.A. and Hörhammer, L. (1973) Vicenin-1 and -2 in the seeds of *Trigonella foenum-graecum* Linn. *Phytochemistry*, 12, 2548.

Wang, D., Sun, H., Han, Y., Wang, X. and Yuan, C. (1997). Studies on chemical constituents of stems and leaves of *Trigonella foenum-graecum* L. *Zhongguo Zhongyao Zazhi*, 22(8), 486–7. CA 128: 306185.

Yoshikawa, M., Murakami, T., Komatsu, H., Murakami, N., Yamahara, J. and Matsuda, H. (1997) Medicinal foodstuffs. IV. Fenugreek seed. (1): Structures of trigoneosides Ia, Ib, IIa, IIb, IIIa, and IIIb, new furostanol saponins from the seeds of Indian *Trigonella foenum-graecum* L. *Chem. Pharm. Bull.*, 45(1), 81–7.

Yoshikawa, M., Murakami, T., Komatsu, H., Yamahara, J. and Matsuda, H. (1998) Medicinal foodstuffs. VIII. Fenugreek seed. (2): Structures of six new furostanol saponins, trigoneosides IVa, Va, VI, Vb, VIIb and VIIIb from the seeds of Indian *Trigonella foenum-graecum* L. *Heterocycles*, 47(1), 397–405.

Zambo, I. and Szilagyi, I. (1982) UV spectrophotometric determination of the Δ^5-steroidal saponin content of *Dioscorea, Trigonella* and *Solanum* species and their tissue cultures. *Herba Hung.*, 21(1–2), 237–44. CA 99: 93818e.

10 Pharmacological properties

Molham Al-Habori and Amala Raman

Introduction

The most widely used species of Trigonella for both medicinal and culinary purposes is *Trigonella foenum-graecum* L., or fenugreek. Fenugreek is an annual plant, extensively cultivated as a food crop in India, the Mediterranean region, North Africa and Yemen. Fenugreek seeds are well known for their pungent aromatic properties (Max, 1992). As a spice, they are a component of many curry preparations (Parry, 1943) and are often used to flavour food and stimulate appetite. Chronic oral administration of ethanolic fenugreek extract (10 mg/day per 300 g body weight) significantly increases food intake and the motivation to eat in rats (Petit *et al.*, 1993), which might be related to the aromatic properties of the seeds (Girardon *et al.*, 1985). Fenugreek seeds are used in India as a condiment, in Egypt as a supplement to wheat and maize flour for bread-making, and in Yemen it is one of the main constituents of the normal daily diet of the general population. Fenugreek leaves are widely consumed in India as a green, leafy vegetable, and are a rich source of calcium, iron, B-carotene and other vitamins (Sharma, 1986b).

Trigonella foenum-graecum L. (in Arabic, *Hulabah*) is also employed as a herbal medicine in many parts of the world. Its leaves are used for their cooling properties and its seeds for their carminative, tonic and aphrodisiac effects (Chopra *et al.*, 1982). It is assumed to have a stimulating effect on the digestive process (Fazli and Hardman, 1968). Fenugreek seeds, which are described in the Greek and Latin Pharmacopoeias, are said to have anti-diabetic activity (Moissides, 1939; Shani *et al.*, 1974; Bever and Zahnd, 1979), and hypocholesterolaemic effects (Singhal *et al.*, 1982; Sharma, 1984). In addition, fenugreek has been reported to possess a curative gastric anti-ulcer action (Al-Meshal *et al.*, 1985), anti-bacterial (Alkofahi *et al.*, 1996), anthelmintic (Ghfaganzi *et al.*, 1980), anti-fertility effects (Setty *et al.*, 1976; Khare *et al.*, 1983; Sethi *et al.*, 1990; Kamal *et al.*, 1993) and anti-nociceptive (Javan *et al.*, 1997) effects.

The aim of this chapter is to review the various pharmacological properties of *Trigonella foenum-graecum*, which appears to be the only species of *Trigonella* with widespread medicinal uses.

Chemical analysis

A chemical analysis of fenugreek indicates that the seeds are a rich source of protein, unavailable carbohydrate, mucilages and saponins (Sauvaire and Baccou, 1976; Baccou *et al.*, 1978; El-Mahdy and El-Sebaiy, 1985; Udayasekhara Rao and Sharma, 1987). Fenugreek resembles Guar (*Cyamopsis tetragonolobus*) in its content of high dietary fibre and high viscosity polysaccharide (Chatterjee *et al.*, 1982; Valette *et al.*, 1984). Fenugreek seeds are also rich in saponins (Sharma, 1986a). Anis and Aminuddin (1985) have reported the presence of three steroidal sapogenins: diosgenin (Figure 10.1), gitogenin and tigogenin. The use of more sophisticated analytical

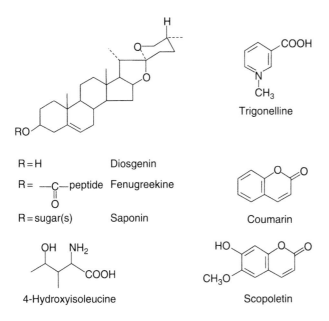

Figure 10.1 Putative anti-diabetic or hypocholesterolaemic compounds in fenugreek seeds.

techniques including gas chromatography coupled with mass spectrometry (GC-MS) has allowed the detection and identification of ten different sapogenins (Brenac and Sauvaire, 1996). The presence of a sapogenin peptide ester, fenugreekine (Figure 10.1) has been reported (Ghosal *et al.*, 1974). More recently, Yoshikawa *et al.* (1997) have isolated six trigoneosides, novel saponins based on furostanol aglycones. Some of the biological properties of the purified steroidal saponins have been evaluated (Sauvaire *et al.*, 1996) and include hypocholesterolaemic and anti-fungal activity as well as effects on food intake, feeding behaviour and motivation in rats (Petit *et al.*, 1995b). Except for differences in fat and saponin content, fenugreek seed powder and defatted fenugreek are chemically similar, containing almost equal amounts of amino acids, minerals and vitamins. Fenugreek like other legumes, is rich in arginine, alanine and glycine, but poor in lysine content (Gopalan *et al.*, 1978; Sharma, 1984). However, 4-hydroxy-isoleucine (Figure 10.1) has been found to be a major free amino acid in the seeds (Sauvaire *et al.*, 1984). Trigonelline (Figure 10.1) is an important alkaloidal component of the seeds (Mishkinsky *et al.*, 1967). The seed contains less starch but higher proportions of minerals (Ca, P, Fe, Zn and Mn) compared with other grain legumes (Sankara Rao and Deosthale, 1981). The total lipid content (7.5 per cent) of the seeds consists of neutral lipids, glycolipids and phospholipids (Hemavathy and Prabhakar, 1989). The aromatic constituents of the seeds have been elucidated (Girardon *et al.*, 1985) and include *n*-alkanes, sesquiterpenes and some oxygenated compounds such as hexanol and γ-nonalactone. The seeds are also known to contain flavonoids, carotenoids, coumarins and other components with very low LD_{50} values (Varshney and Sharma, 1996).

Anti-diabetic effects

Diabetes mellitus (DM) is a wide-spread disorder that has long been recognised in the history of medicine (Best, 1962; West, 1978). Before the advent of insulin and oral hypoglycaemic drugs the major form of treatment involved the use of plants. More than 400 plants are known to have been recommended traditionally, and recent investigations have affirmed the potential value of some of these treatments (Marles and Farnsworth, 1995; Bailey and Day, 1989). The hypoglycaemic and/or anti-hyperglycaemic effect of several plants used as anti-diabetic remedies has been confirmed and the mechanisms of their activity are being studied (Marles and Farnsworth, 1995). Chemical studies directed at the isolation, purification and identification of the substances responsible for the anti-diabetic activity, are also being conducted (Attaur-Rahman and Zaman, 1989; Bailey and Day, 1989; Ivorra *et al.*, 1989; Alacron-Aguilar *et al.*, 1993; Sadhukhan *et al.*, 1994; Marles and Farnsworth, 1995).

Fenugreek seeds have been known for a long time for their anti-diabetic action (Moissides, 1939; Mishkinsky *et al.*, 1967). Fourier (1948) observed that the consumption of coarsely ground fenugreek seeds improved severe diabetes in human subjects. This property was later confirmed in alloxan-diabetic rats, where the seed extract induced a significant hypoglycaemic effect (Bever and Zahnd, 1979; Khosla *et al.*, 1995a), as did its major alkaloid, trigonelline (Shani *et al.*, 1974). Ghafghazi *et al.* (1977) have shown that an extract of fenugreek prevented the hyperglycaemia induced by cadmium and alloxan in rats. Amin *et al.* (1988) also showed that diabetic animals that were treated with a 20 per cent fenugreek diet 5 weeks prior to a streptozotocin (STZ) injection, showed a general improvement in clinical status compared to animals treated with STZ alone. Hyperglycaemia, free fatty acids, cholesterol and triglycerides were significantly reduced. However, if the pretreatment period was not used, a supplementary diet of fenugreek following the induction of diabetes did not improve the diabetic state, as judged by blood glucose and lipid levels. Thus a possible preventive role for fenugreek against chemically induced diabetes has been suggested.

A beneficial effect in pre-existing diabetic states has, however, also been shown in numerous other studies (Table 10.1). A reduction in hyperglycaemia was observed in diabetic dogs fed with fenugreek seeds (Ribes *et al.*, 1984; 1986), and in mice where 40–80 per cent dilution of a fenugreek decoction and an ethanolic extract (200–400 mg/kg) were used (Ajabnoor and Tilmisany, 1988). Similar effects were reported in healthy human volunteers given fenugreek powder (25 g/day) mixed in their diet (Sharma, 1986b): Type I diabetics were fed fenugreek (100 g/day) (Sharma *et al.*, 1990) and Type II diabetics were fed fenugreek (15 g/day) (Madar *et al.*, 1988; Sharma and Raghuram, 1990). Fenugreek seeds (whole as well as extracted) were found to diminish hyperglycaemia in normal and diabetic subjects (Sharma, 1986b; Sharma *et al.*, 1990). Fasting blood glucose, 24-h urinary sugar excretion and serum cholesterol were also significantly reduced in these subjects.

Despite a significant reduction in postprandial glucose, in some studies no significant change was observed in plasma insulin following fenugreek administration to non-insulin dependent 'NIDDM' diabetics (Madar *et al.*, 1988), rats (Madar, 1984), or dogs (Ribes *et al.*, 1984; 1986). However, other studies in chemically-induced diabetic rats have demonstrated a significant increase in plasma insulin levels (Sharma, 1986b; Petit *et al.*, 1993; 1995a). These conflicting results may be due to differences in the type of fenugreek preparation used in the various studies (Table 10.1). The observed increase in plasma insulin levels following administration of an ethanolic fenugreek extract to rats (Petit *et al.*, 1993) was suggested to be due either to a direct stimulatory effect on the β-cells or to an indirect effect related to the palatability and the flavour-enhancer properties of the extract. The latter hypothesis was put forward in line with the effect of the sweet taste of saccharin solution, which has been reported to trigger a rapid cephalic phase of insulin response in the absence of any significant change in glycaemia (Berthoud *et al.*, 1981). However, the presence in fenugreek of an insulin-secretion stimulating compound (4-hydroxyisoleucine) has also been reported (Hillaire-Buys *et al.*, 1993; Petit *et al.*, 1995a; Sauvaire *et al.*, 1996).

Table 10.1 Summary of the reported anti-diabetic properties of fenugreek *in vivo*

Test substance	Administered to	Dose	Effects observed	References
Fenugreek powder	Non-diabetic rats	2–8 g/kg for 2 weeks	Hypoglycaemic	1
		20% of diet for 2 weeks	Anti-hyperglycaemic	2
		250 mg (single dose)	Anti-hyperglycaemic	3
	Diabetic rats	2–8 g/kg for 2 weeks	Hypoglycaemic	1
	Non-diabetic humans	10 g per day for 2 days	No effect on OGTT	4
		25 g (single dose)	Anti-hyperglycaemic	5
	NIDDM humans	10 g per day for 2 days	Anti-hyperglycaemic	4
		15 g per day for 4–7 days	Anti-hyperglycaemic/ no increase in plasma insulin	6
		25 g per day for 15 days	Anti-hyperglycaemic against intravenous GTT	7
		25 g per day for 3 weeks	Anti-hyperglycaemic	5
	IDDM humans	100 g per day for 10 days	Hypoglycaemic and anti-hyperglycaemic	8
Suspension	Non-diabetic rats	25 g per 5 ml	No effect on OGTT	9
	Diabetic rats	0.25 g per 5 ml	Anti-hyperglycaemic	9
Decoction	Diabetic & non-diabetic rats	40–80% dilution	Anti-hyperglycaemic	10
Oil fraction	Diabetic & non-diabetic	Corresponding to 7% of whole fenugreek seeds	No effect on blood glucose	11,12
Defatted fraction	Non-diabetic dogs	Corresponding to 93% of whole fenugreek seeds	No effect on blood glucose	12
	Diabetic dogs	Corresponding to 93% of whole fenugreek seeds	Hypoglycaemic anti-hyperglycaemic	11,12
	NIDDM humans	25 g per day for 3 weeks	Anti-hyperglycaemic	5
Defatted subfractions 'a' (fibre)	Diabetic dogs	Amount corresponding to total defatted fraction fed for 3 weeks	Anti-hyperglycaemic	13,14
'b' (protein + saponin)		As above	No effect on OGTT	13,14
'P' (protein)		As above	No effect on OGTT	14
'S' (saponin)		As above	No effect on OGTT	14
Ethanolic extract	Diabetic & non-diabetic rats	200–400 mg/kg	Anti-hyperglycaemic	10
	Non-diabetic	250 mg/kg	Anti-hyperglycaemic	3
	Diabetic rats	5 mg/kg for 3 weeks	Anti-hyperglycaemic	15

Note
1. Khosla *et al.*, 1995a; 2. Amin *et al.*, 1987; 3. Ali *et al.*, 1995; 4. Sadhukhan *et al.*, 1994; 5. Sharma, 1986b; 6. Madar *et al.*, 1988; 7. Raghuram *et al.*, 1994; 8. Sharma *et al.*, 1990; 9. Madar, 1984; 10. Ajabnoor and Tilmisany, 1980; 11. Ribes *et al.*, 1984; 12. Valette *et al.*, 1984; 13. Ribes *et al.*, 1986; 14. Ribes *et al.*, 1987; 15. Shani *et al.*, 1974.

Apart from biochemical improvements, fenugreek seeds have been reported to markedly suppress the clinical symptoms of diabetes such as polyuria, polydypsia, weakness and weight losses (Sharma, 1986b). It has also been demonstrated that the hypoglycaemic property of fenugreek is not destroyed by the cooking or roasting process (Sharma, 1986b; Khosla *et al.*, 1995a).

A number of investigations have been carried out to identify the factors responsible for the anti-diabetic activity of fenugreek and the mechanisms involved in this effect. One group of researchers have studied two fractions of the seed, namely, the lipid extract and the defatted seed material that contains fibres, saponins and proteins (Ribes *et al.*, 1984; Valette *et al.*, 1984). The above work led to the suggestion that the active component was not in the lipid extract but in the defatted portion of the seeds, which provoked a decrease in hyperglycaemia and hypercholesterolaemia in both normal and diabetic dogs. Defatted fenugreek had an influence on the response to oral glucose tolerance test (OGTT) and modified not only the blood glucose level but also pancreatic hormone levels (Ribes *et al.*, 1984; 1986). It decreased the normally observed peak plasma insulin levels in normal dogs following OGTT (Ribes *et al.*, 1984), as well as levels of glucagon (an aggravating factor of diabetes) and somatostatin (observed after OGTT) in diabetic dogs (Ribes *et al.*, 1986), which infers better carbohydrate regulation. This defatted fraction was further investigated (Ribes *et al.*, 1986) by dividing it into two subfractions: subfraction 'a' that contained the testa and endosperm and is rich in fibre (79.6 per cent), and subfraction 'b' that contained the cotyledons and axles and is rich in proteins (52.8 per cent) and saponins (7.2 per cent). Their results, like those of Madar (1984) and Sharma (1986b), showed that the anti-diabetic property of fenugreek seeds was contained in the testa and endosperm subfraction. The authors suggest that although rich in fibres, it is not possible to exclude the coexistence of one or more unknown, pharmacologically-active compounds in this subfraction of the seed.

In early reports, the hypoglycaemic effect of fenugreek was attributed to its major alkaloid, trigonelline (Mishkinsky, 1967; Shani *et al.*, 1974). Trigonelline (Figure 10.1) is the N-methyl derivative of the vitamin nicotinic acid, and is excreted in human and rat urine after oral administration of nicotinic acid (Ackerman, 1912), but when fed to cats, dogs and rabbits it is excreted unchanged (Kohlrausch, 1912). However administration of trigonelline, in the amounts present in fenugreek, to diabetic patients did not show any significant hypoglycaemic activity (National Institute of Nutrition, 1987). Furthermore, a recently isolated active hypoglycaemic principle from fenugreek has been shown to be different from trigonelline (Moorthy *et al.*, 1989). Moorthy *et al.* (1989) reported the presence of an orally active principle isolated from fenugreek seeds, which improves glucose tolerance for a period of 1 week in alloxan-treated rabbits. This fraction, which was different from and more potent than trigonelline, was also reported to decrease fasting blood glucose in alloxan-recovered rabbits with an initial fasting blood glucose level of 180 mg/dL. Following daily treatment with this fraction (50 mg/kg) for 1 month, fasting blood glucose decreased by about 50 per cent in severely diabetic rabbits with an initial fasting blood glucose of 400 mg/dL. In addition, there was an improvement in glycosylated haemoglobin and serum lipid profile, an increase in the activity of key glycolytic enzymes in muscle but not in the liver and a slight, though not statistically significant, inhibition of key gluconeogenic enzymes in the liver and kidney. However, no reports were found on the chemical composition of this active fraction.

In 1993, Hillaire-Buys *et al.* reported the presence of an insulin-stimulating substance in the seeds of fenugreek. This compound was obtained by sequential chromatography from defatted fenugreek seeds and identified as 4-hydroxyisoleucine (Figure 10.1). 4-Hydroxyisoleucine (200 µmol/L) evoked a biphasic insulin response *in vitro*, using isolated pancreas perfused with glucose (Petit *et al.*, 1995a; Sauvaire *et al.*, 1996). This response increased in a concentration dependent manner both *in vitro* and *in vivo* in conscious fasted dogs. It was effective after oral administration and improved oral glucose tolerance (Sauvaire *et al.*, 1996). The data showed 4-hydroxyisoleucine, which represents up to 80 per cent of free amino acids in fenugreek seeds (Sauvaire *et al.*, 1984), to stimulate insulin secretion only in the presence of intermediate to high glucose concentrations and to be effective in a much lower concentration range than its structural amino acid congeners leucine and isoleucine. The isolated 4-hyroxyisoleucine was found to partially affect the K^+-conductance of the β-cell plasma membrane. 4-Hydroxyisoleucine is an unusual amino acid that was isolated and identified for the first time by Fowden *et al.* (1973), its conformation was established by Alcock *et al.* (1989).

Other postulated hypoglycaemic constituents of fenugreek (Figure 10.1) are coumarin, which was shown to have a profound hypoglycaemic effect in normal and alloxan-induced diabetic rats (Shani *et al.*, 1974), scopoletin, another coumarin constituent that exerted borderline hypoglycaemic effects in normal and alloxan-induced diabetic rats (Shani *et al.*, 1974), and fenugreekine, a peptide ester of diosgenin and one or more units of 4-hydroxyisoleucine. Fenugreekine is stated to have a hypoglycaemic effect although details are not given (Ghosal *et al.*, 1974). The relationship of hypoglycaemic doses of these compounds to their concentration in active fenugreek preparations needs further exploration.

The endosperm of the fenugreek seed is a rich source of fibre (20 per cent) and gum (32.4 per cent) (Sharma, 1986b). It is known that the addition of fibre to the diet of diabetics results in a reduction of blood glucose during OGTT (Jenkins *et al.*, 1978; Jenkins, 1979; Monnier *et al.*, 1978). The clinical role of dietary fibre in glycaemic control has been reviewed (Jenkins and Jenkins, 1984; Vinik and Jenkins, 1988). Furthermore a high viscosity of gut contents has been reported to inhibit the intestinal absorption of glucose (Johnson and Gee, 1980) and significantly reduce the mean postprandial blood glucose and insulin curve (O'Connor *et al.*, 1981). This effect has been attributed, for example, to the viscosity of hydrated guar gum, which reduces the rate of gastric emptying (Holt *et al.*, 1979; Blackburn *et al.*, 1984). Fenugreek, like guar gum, is very viscous and is rich in galactomannan (Reid and Meier, 1970).

In view of its high content of soluble fibre, it has been postulated that one mechanism by which fenugreek may modulate plasma glucose levels is by delaying gastric emptying and by direct interference with glucose absorption at the gastrointestinal level (Madar, 1984). The latter effect was investigated *in vitro* using inverted gut sac from the jejunum of male rats, where the addition of 0.1–1 per cent fenugreek seed powder to the mucosal side significantly inhibited the 3-0-methyl-D-glucose transport into the serosal side (Madar, 1984). Based on the finding that whole fenugreek seeds, extracted fenugreek seeds and gum isolate are rich sources of fibre in the form of galactomannan (Sharma, 1986b), which resembles guar gum in chemical structure and viscosity (16–20 cP) (Ribes *et al.*, 1984), it was concluded that the dietary fibre in fenugreek is the major contributor in reducing plasma glucose (Sharma, 1986b; Madar *et al.*, 1988). Furthermore, the fact that fenugreek had no significant effect on insulin levels in these studies suggested that it decreased glucose levels by inhibition of diffusion, or transport of glucose without involvement of intestinal hormonal factor (Madar *et al.*, 1988). Degummed fenugreek seed was shown to have little hypoglycaemic effect, further excluding non-mucilagenous fibre as the cause of the effect observed (Sharma, 1986b). It has recently been shown that galactomannan, in the gel fraction of the seeds, is a factor which reduces the plasma glucose in both *in vivo* and *in vitro* studies using inverted gut, by increasing the viscosity of the gut contents (Madar and Shomer, 1990).

In more recent studies, Ali *et al.* (1995) showed that fenugreek powder, its methanolic extract, and the residue remaining after methanol extraction all had significant anti-hyperglycaemic effects when fed simultaneously with glucose. The soluble dietary fibre (SDF) fraction showed no effect on the fasting blood glucose levels of non-diabetic or NIDDM model rats. However, when fed simultaneously with glucose, it showed a significant anti-hyperglycaemic effect in NIDDM model rats suggesting that fibre might be responsible for the observed improvement in the glucose tolerance but did not contribute to the hypoglycaemic effects. Thus other mechanisms and components may be associated with the decrease in basal glycaemia following fenugreek administration, which has been observed in some studies (Table 10.1).

Fenugreek has an additional possible mode of action: it has an inhibitory effect on intestinal carbohydrate digestion. Fenugreek was found to decrease digestion of starch and also glucose absorption both *in vivo* (by following a tolerance test of a meal containing starch) and *in vitro* using the inverted sac technique (Madar and Shomer, 1990). This may be the result of a direct inhibitory effect on the digestive enzymes or of reduced enzyme–substrate contact (Wong *et al.*, 1985; Edwards *et al.*, 1988). Amin *et al.* (1987) demonstrated the existence of a low relative molecular mass fraction in the aqueous extract of fenugreek that inhibits carbohydrate degrading enzymes (α-amylase and sucrase) in rat intestines. These results are in line with earlier reports which observed that inhibiting intestinal disaccharidase activities by acarbose moderated the development of diabetes in STZ-treated rats (Goda *et al.*, 1982). Recently, Platel and Srinivasan (1996) reported a significant decrease in the activity of intestinal sucrase with the addition of 2 per cent fenugreek seeds to the diet of rats, with very little effect on α-amylase, maltase and lactase.

The above studies suggest that fibre and other components in fenugreek seed, acting at the gastrointestinal level, may well be responsible for the observed improvement in oral glucose and starch tolerance. However, this does not explain the hypoglycaemic effects (reduction in basal glycaemia) observed in some studies, and other mechanisms are possible. Ajabnoor and Tilmisany (1988) using both a 40–80 per cent dilution of fenugreek decoction and an ethanol extract (200–400 mg/kg) of the seeds in normal and alloxan-treated diabetic male albino mice, further confirmed the earlier reports of hypoglycaemic effects and put forward the argument that since their experiments were conducted on fasting mice, the effect could not be due to the gastrointestinal action of fibre. The authors went further to suggest that the mechanism of antidiabetic action of the seeds may be similar to that of tolbutamide, although other mechanisms are possible. Moreover, Raghuram *et al.* (1994) showed that fenugreek powder (25 g) when given in the diet for 15 days to NIDDM patients prior to an intravenous glucose load, significantly altered plasma glucose kinetics, reducing the area under the plasma glucose curve and increasing the metabolic clearance rate. In addition, fenugreek increased the molar insulin-binding sites on erythrocytes. However, serum insulin levels were not measured (Raghuram *et al.*, 1994). This study suggests that fenugreek can improve peripheral glucose utilisation and that it may exert its anti-diabetic activity by effects at the insulin receptor as well as at the gastrointestinal level.

Thus the hypoglycaemic and anti-hyperglycaemic actions of fenugreek have been attributed both to the gastrointestinal effects of local dietary fibre (Madar, 1984) and to systemic effects of active principles, such as 4-hydroxyisoleucine, present in the seeds (Ribes *et al.*, 1986; Moorthy *et al.*, 1989; Hillaire-Buys *et al.*, 1993; Sauvaire *et al.*, 1996). Trigonelline has been discounted as an active principle by more recent studies (National Institute of Nutrition, 1987), while claims for the activity of fenugreekine (Ghosal *et al.*, 1974) remain unsubstantiated.

The studies reported so far in this section have examined the anti-diabetic effects of fenugreek seeds. By contrast, Abdel-Barry *et al.* (1997) reported that an aqueous extract of *Trigonella foenum-graecum* leaves (0.5–1 g/kg) could lower basal glycaemia on intra-peritoneal administration to normoglycaemic and hyperglycaemic (alloxan-treated) rats. Oral administration of the aqueous extract (1–8 g/kg) and intra-peritoneal administration of an ethanolic extract (0.8 g/kg) decreased glycaemia in hyperglycaemic but not normoglycaemic rats.

Hypocholesterolaemic effects

The association of raised serum cholesterol with cardiovascular disease is well known (Gordon *et al.*, 1977). Some studies suggest that elevated serum triglyceride may also be a risk factor (Carlson *et al.*, 1979; Carlson and Bottiger, 1985) especially in individuals with diabetes

(West *et al.*, 1983), there is often a marked hyperlipidaemia in diabetes (Maison and Boucher, 1978; Betteridge, 1989). Moreover, diabetic patients experience a two- to three-fold increase in cardiovascular morbidity and mortality when compared with non-diabetics. The beneficial effect of lowering elevated serum cholesterol levels for the prevention of coronary heart disease (CHD) has been well established (Lipid Research Clinics Program, 1984). Dietary intervention has been recommended for all subjects with a low density lipoprotein (LDL) level of more than 160 mg/dL (Report of the National Cholesterol Education Program, 1988). In addition to the quantity of fat and the polyunsaturated/saturated fat ratio, other dietary factors also play a role in the management of hyperlipidaemia (Grundy, 1987). Several studies have shown that dietary fibre, particularly soluble fibre, has considerable influence on serum cholesterol levels (Kritchevsky, 1982; Dreher, 1987; Miettinen, 1987).

Research carried out on legumes has led to the belief that they are beneficial in lowering the total cholesterol levels in humans (Madar and Odes, 1990; Sharma *et al.*, 1990; Sharma *et al.*, 1996a). Scientific reports indicate that fenugreek does indeed have therapeutic properties that may be beneficial in treating hypercholesterolaemia (Table 10.2). Fenugreek seeds have been shown to possess a hypocholesterolaemic effect in rats (Singhal *et al.*, 1982; Sharma, 1984; 1986a; Stark and Madar, 1993; Khosla *et al.*, 1995b) and dogs (Valette *et al.*, 1984). Elevation of cholesterol levels in the rat was prevented by adding fenugreek at 15–60 per cent to a hypercholesterolaemia-inducing diet (Sharma, 1984). Fenugreek was demonstrated to have a greater effect on exogenous cholesterol (when given with a hypercholesterolaemia-inducing diet containing 1 per cent cholesterol) than on endogenous cholesterol (fenugreek given with a cholesterol-free stock diet) (Sharma, 1984). Defatted fenugreek (100 g) incorporated in the experimental diet of hyperlipidaemic non-diabetic subjects significantly reduced serum total cholesterol, LDL and very low density lipoprotein (VLDL)-cholesterol and triglyceride levels (Sharma *et al.*, 1991), with no observed changes in high density lipoprotein (HDL)-cholesterol. As a result, there was a significant increase in the ratio of HDL to total cholesterol and HDL to that of LDL and VLDL-cholesterol, which have been shown to be reliable risk assessment factors of CHD (Kannel, 1983).

In a short-term study, fenugreek seeds were also found to exert hypocholesterolaemic activity in diabetic patients (Sharma and Raghuram, 1990; Sharma *et al.*, 1990). In NIDDM patients, ingestion of an experimental diet containing 25 g fenugreek seed powder for 24 weeks resulted in a significant reduction of total cholesterol, LDL- and VLDL-cholesterol and triglyceride levels (Sharma *et al.*, 1996a). Serum cholesterol was significantly reduced and this fall was mainly due to a reduction in LDL and VLDL fractions. Triglyceride levels also showed a similar change. On the other hand, HDL-cholesterol showed a slight rise ($P > 0.05$). The overall results are in agreement with earlier observations made in diabetic patients (Sharma, 1986a; Sharma *et al.*, 1990). All the lipid parameters improved rapidly during the initial 8 weeks after the incorporation of fenugreek with a slower change thereafter (Sharma *et al.*, 1996a). An increase in HDL-cholesterol was also observed in diabetic rats fed 2–8 g/kg body weight of unroasted and roasted fenugreek seeds for 2 weeks (Khosla *et al.*, 1995b). These results indicate a potential beneficial effect of fenugreek seeds in the lipid profile of diabetic subjects, in addition to the effects on glycaemia reviewed earlier.

The ability of fenugreek to selectively reduce the LDL and VLDL fraction of total cholesterol could be beneficial in preventing atherosclerosis. A similar selective effect on LDL-cholesterol was observed with dietary fibres such as oat bran (Kirby *et al.*, 1981) and guar gum (Jenkins *et al.*, 1980). Natural carbohydrates rich in fibre content have been found to be effective against hyperlipidaemia and ischaemic heart disease (Trowell, 1972). Insulin secretion has been shown to regulate VLDL and triglyceride concentration (Sparks and Sparks, 1994), the hormone has

Table 10.2 Summary of the reported hypocholesterolaemic and hypolipidaemic effects of fenugreek *in vivo*

Test substance	Administered to	Dose	Effects observed	References
Fenugreek powder	Normal rats	2–8 g/kg for 2 weeks	Decrease in plasma cholesterol, triglyceride, VLDL- and LDL-cholesterol	1
		50% of diet for 2 weeks	Decrease in plasma cholesterol	2
	Diabetic rats	2–8 g/kg for 2 weeks	Decrease in plasma cholesterol and triglyceride	1
	Hypercholesterolaemic rats	50% of diet for 2 weeks	Decrease in plasma cholesterol	2
		10–60% of diet for 4–6 weeks	Decrease in plasma cholesterol, VLDL- and LDL-cholesterol	3
		30% of diet for 4 weeks	Decrease in plasma cholesterol	4
	NIDDM humans	25 g per day for 24 weeks	Decrease in plasma cholesterol, triglyceride, VLDL- and LDL-cholesterol	5
		15 g per day for 4–7 days	No effect on plasma lipids following a meal tolerance test	6
	IDDM humans	100 g per day for 10 days	Decrease in plasma cholesterol and triglyceride	7
Oil fraction	Diabetic and non-diabetic dogs	Corresponding to 7% of whole fenugreek seeds	No effect on plasma cholesterol	8
Defatted fraction	Diabetic and non-diabetic dogs	Corresponding to 93% of whole fenugreek seeds for 3 days	Decrease in plasma cholesterol	8
	Hyperlipidaemic subjects	100 g defatted fenugreek for 20 days	Decrease in plasma cholesterol, triglyceride, VLDL- and LDL-cholesterol	7
Defatted subfractions	Diabetic dogs			
'a' (fibre)		Amount corresponding to total defatted fraction for 3 weeks	Decrease in plasma cholesterol	9
'b' (protein+ saponin)		As above	Decrease in plasma cholesterol and triglyceride	9
'P' (protein)		As above	No effect on plasma lipids	9
'S' (saponin)		As above	Decrease in plasma cholesterol and triglyceride	9
Ethanolic extract	Normal rats	10 mg per day for 2 weeks	Decrease in plasma cholesterol, LDL- and VLDL-cholesterol and increase in plasma insulin	10
		30 g/kg for 4 weeks	Decrease in fasting plasma cholesterol	11

Note

1. Khosla *et al.*, 1995b; 2. Singhal *et al.*, 1982; 3. Sharma, 1984; 4. Sharma, 1986a; 5. Sharma *et al.*, 1996a; 6. Madar *et al.*, 1988; 7. Sharma *et al.*, 1990; 8. Valette *et al.*, 1984; 9. Ribes *et al.*, 1987; 10. Petit *et al.*, 1993; 11. Stark and Madar, 1993.

been found (Bhathena *et al.*, 1974) to stimulate the hepatic production of VLDL. Based on this, a high fibre diet which reduces insulin secretion was used in the treatment of hyperlipidaemia in diabetic subjects (Paisey *et al.*, 1984). Thus the alterations in lipid profiles observed after ingestion of fenugreek, which contains dietary fibre, may have been due to a decreased synthesis of VLDL in the liver. However, since ingestion of fenugreek extracts was reported to stimulate insulin secretion in diabetic rats (Sharma, 1986b; Petit *et al.*, 1993; 1995a) the intermediary role of insulin in altering lipid profiles is unclear.

Among the fenugreek fractions, the lipid extract and 0.12 per cent trigonelline had no hypocholesterolaemic effect (Valette *et al.*, 1984) while the defatted fractions, gum isolate and the crude saponins, fed to normal and diabetic rats at equivalent amounts to that present in a diet containing 30 per cent fenugreek seeds, showed hypocholesterolaemic activity without any significant effect on the triglyceride level (Sharma, 1986a). Further studies by Ribes *et al.* (1987) showed that although subfraction 'a' (79.6 per cent fibre) displays both an anti-diabetic and hypocholesterolaemic activity, subfraction 'b' (52.8 per cent proteins and 7.2 per cent saponins) has a clear hypolipidaemic effect since it reduces elevated cholesterol and triglyceride levels in diabetic dogs. This latter subfraction was further subdivided to two fractions 'S' which contained the saponins (22.2 per cent) and subfraction 'P' containing the totality of the proteins (70.5 per cent). Administration of subfraction 'P', had no effect on the high levels of cholesterol and triglycerides in diabetic dogs. This conclusion is in accordance with that of Sharma (1984), demonstrating that the active principle was not related to the amino acids, and rules out the possibility that alterations in serum cholesterol by fenugreek are related to changes in the lysine/arginine ratio (Kritchevsky *et al.*, 1978). By contrast, the presence of saponins seem essential for the hypolipidaemic activity of fenugreek seeds (Ribes *et al.*, 1987; Sauvaire *et al.*, 1991).

Saponins are plant glycosides whose aglycone structure is triterpenoid or steroidal. They are a heterogeneous group of amphiphilic compounds and are highly surface-active. Most saponins are haemolytic, can bind cholesterol and form stable foams (Price *et al.*, 1987). Studies reported so far on the effects of saponins on cholesterol homeostasis concern mainly the triterpenoid saponins from lucerne (Malinow, 1984) and the steroidal saponin from soya bean (Sidhu *et al.*, 1987; Calvert *et al.*, 1981), which reduce the intestinal uptake of cholesterol. It has also been reported that a steroidal saponin, digitonin, prevents or lowers hypercholesterolaemia in monkeys (Malinow *et al.*, 1978; Oakenfull and Fenwick, 1978) without modifying HDL-cholesterol levels (Malinow *et al.*, 1981). In contrast, Gibney *et al.* (1982) reported no effect of a commercial saponin when fed to rats and hamsters. However, this study mentioned neither the chemical structure nor the origin of the saponin used.

Saponins derived from lucerne (*Medicago sativa,* alfalfa) were found to reduce plasma cholesterol levels by the direct binding of dietary saponins with cholesterol in the digestive tract with subsequent excretion of the complex in the faeces (Malinow *et al.*, 1977; 1981; Story *et al.*, 1984). Other types of saponins affect cholesterol metabolism indirectly by interacting with bile acids and increasing their faecal excretion (Oakenfull *et al.*, 1984). However, whereas lucerne saponins interact directly with cholesterol (Gestetner *et al.*, 1971), soya bean saponins do not appear to do so (Birk, 1969). The results of Stark and Madar (1993) indicate that saponins present in fenugreek, similar to soya bean saponins, do not interact directly with cholesterol. However, using the inverted sac technique, an ethanol extract of fenugreek exhibited a strong inhibitory effect on bile salt absorption (Stark and Madar, 1993), in a quantitative manner. These findings are in agreement with those of Bhat *et al.* (1985) and Sharma (1984), where fenugreek enriched diets were found to increase both faecal weight and excretion of bile acids. The mechanism that causes this effect is still not clear. One possibility, is that large mixed micelles are formed containing bile salts and saponins, and as these large molecules are not available for absorption (Sidhu and Oakenfull, 1986), they are lost in the faeces. The observed lowering of blood and hepatic cholesterol may be due to a subsequent increase in the conversion of cholesterol to bile acids by the liver.

Fenugreek seed saponins are of a steroidal nature with diosgenin (Figure 10.1) as the main sapogenin (Mahato *et al.*, 1982). Diosgenin has various effects on cholesterol metabolism, one of the most important being the capacity to lower plasma cholesterol concentration in chickens and rabbits fed with cholesterol (Laguna *et al.*, 1962). The hypocholesterolaemic effect of diosgenin has been suggested to depend on its capacity to inhibit cholesterol absorption, increase biliary cholesterol secretion, increase faecal excretion of neutral sterols and thus to decrease liver cholesterol concentrations (Cayen and Dvornik, 1979; Uchida *et al.*, 1984; Ulloa and Nervi, 1985).

Malinow (1985) has shown that diosgenin glucoside was more efficient than diosgenin in reducing intestinal absorption of cholesterol. At comparable small doses, diosgenin glucoside inhibited cholesterol absorption *in vivo* and *in vitro*, whereas diosgenin did not (Malinow, 1985; Malinow *et al.*, 1987). Sauvaire *et al.* (1991) have examined the transformation of fenugreek subfractions rich in steroid saponins during their passage through the digestive tract, to determine the relative contribution of saponins and/or diosgenin and other steroid sapogenins to the hypocholesterolaemic effect of fenugreek seeds. Faecal samples from alloxan diabetic dogs fed with the fenugreek subfractions were analyzed by capillary GC-MS for the presence of sapogenins. The results suggest that saponins, are in part (about 57 per cent), hydrolysed into sapogenins (disogenin, smilagenin, gitogenin) in the digestive tract, the location of fenugreek saponin hydrolysis in the digestive tract was not determined. The authors concluded that saponin hydrolysis does occur, presumably in the stomach and/or in the proximal small intestine (Sauvaire *et al.*, 1991), but since hydrolysis was incomplete, saponins may be implicated, alone or together with sapogenin, in the observed hypocholesterolaemic effect of fenugreek seeds.

Apart from the role of fenugreek saponins and sapogenin, it has been suggested that the inhibition of bile salt absorption may be primarily mechanical, due to the formation of a physical barrier by fenugreek extracts such as the gel fraction. A study by Ribes *et al.* (1987) showed that a fibre-rich subfraction ('a') separated from the saponins also displayed a hypocholesterolaemic effect. Galactomannan derived from fenugreek seeds has been reported to inhibit intestinal bile acid absorption, reducing the efficiency of their enterohepatic circulation and subsequently decreasing plasma cholesterol level (Madar and Shomer, 1990).

Anti-fertility effects

Efforts have been made to study the contraceptive and anti-fertility effects of crude extracts of plants of a diverse nature (Rao *et al.*, 1988; Sethi *et al.*, 1990; Desta, 1994), but as yet not a single plant has been found to be successful as a potent clinically effective contraceptive agent. A number of studies have been conducted on the potential use of fenugreek in contraception.

Effects in the male

Fenugreek has been used as a spermicidal agent in albino rats (Dhawan *et al.*, 1977) and in *in vitro* studies utilizing human semen (Setty *et al.*, 1976). The *n*-butanol extract of fenugreek at 2 per cent has been reported to have spermicidal activity; this has been related to the saponins present in this fraction (Setty *et al.*, 1976). Further studies of saponins of known chemical structure revealed that the spermicidal potency is associated with β-amyrin C-28 carboxylic acid type of sapogenin(s) such as hederagenin, oleanolic and basic acids. β-Amyrin sapogenins without C-28 carboxylic acid such as glycyrrhetic acid, and bacogenin or the

α-amyrin C-28 carboxylic acid type of sapogenins such as brahmic acid and asiatic acid are devoid of sperm immobilizing properties (Setty *et al.*, 1976). Moreover, the degree of activity of a particular saponin appears to be dependent upon a specific sequence of attachment of sugar moieties on certain genin molecules (hederagenin, oleanolic or basic acids) besides the presence of a free C-28 carboxylic acid group in the β-amyrin nucleus (Stolzenberg and Parkhurst, 1974; Setty *et al.*, 1976).

Kamal *et al.* (1993) reported that the steroidal extract (sapogenin) of fenugreek seeds at 100 mg/day/rat administered orally for 60 days significantly reduced the weight of the testis epididymis, ventral prostate and seminal vesicles, with no differences in the body weight with respect to the control group. A fertility test gave 100 per cent negative results in the treated group, whereas libido remained unchanged as evidenced by the vaginal plugs in females kept in the same cage (Kamal *et al.*, 1993). This loss of fertility was attributed to decreased spermatozoal density and motility of cauda epididymis. The reduction in the reproductive organs' weight may indicate a decrease in the circulating levels of androgen (Chinoy *et al.*, 1982). A decline in other androgen dependent parameters, that is, protein, sialic acid and fructose, also suggest a reduction in androgen levels (Prasad and Rajlakshmi, 1976; Kamal *et al.*, 1993). Thus fenugreek extracts may exert anti-fertility and anti-androgenic activity in male albino rats (Kamal *et al.*, 1993).

Effects in the female

While consumption of fenugreek seeds by women during lactation is highly recommended in India (Nadkarni, 1954), its use in pregnancy is restricted. Studies in female rats fed diets containing 5 or 20 per cent fenugreek seed powder for a period of 21 days (Mital and Gopaldas, 1986a) showed no significant effect on the number of implantations, number of resorptions, or foetal and placental weight as compared to the control groups. Another study, where fenugreek-treated rats were allowed to continue pregnancy to full term and give birth, showed no significant effect on the litter size (Mital and Gopaldas, 1986b). In addition, the findings of Mital and Gopaldas (1986b) demonstrated no additional beneficial effect of fenugreek seeds during the lactation period contrary to an earlier study of El-Ridi *et al.* (1954), which suggested that the oil extracted from the fenugreek seeds contained a lactation promoting factor.

In contrast to the above studies, Khare *et al.* (1983) reported a mild anti-fertility effect of feeding an ethereal extract of fenugreek seeds to female rats, where the absence of foetal implants was regarded as an anti-fertility index. It has been postulated that the ethereal extract is a concentrated source of the steroidal substance diosgenin (Figure 10.1), which is used as a starting material in the synthesis of sex hormones and oral contraceptives (Shankaracharya and Natarajan, 1972). The dose administered was 25 mg of extract per 100 g body weight. Based on the fact that the ethereal extract or 'oil fraction' is 7 per cent of the whole fenugreek seed powder, the 25 mg of the ethereal extract used by Khare *et al.* (1983) is equivalent to 357 mg of fenugreek seeds, which appears to be lower than that used by Mital and Gopaldas (1986a).

Fenugreek seed powder at 175 mg/kg administered daily to mature adult female albino rats for the first 10 days of the post-mating period showed an 18 per cent abortifacient activity compared with the 2 per cent seen in the control group (Sethi *et al.*, 1990). In the same study the number of resorptions was 10 compared with 1 in the control. A more recent study by Elbetieha *et al.* (1996) showed that the aqueous extract of fenugreek administered orally by intragastric intubations to female rats for the first 6 days of pregnancy did not produce effects significantly different from the control group. There was, however, a 66 per cent increase in the number of resorptions in those females treated with fenugreek (Elbetieha *et al.*, 1996). Embryonic resorption most probably resulted from the transplacental passage of the substance to the embryo in addition to modification of the uterine lining function. The effect of fenugreek was attributed to a possible estrogenic activity (Al-Hamood and Al-Bayatti, 1995). It is well known that estrogens, depending on the dose, are responsible for stimulating uterine contractility and restricting the development of implanted embryo.

Miscellaneous pharmacological effects

Gastric ulcer and wound healing effects

Fenugreek, in the form of a tea, is used as a herbal remedy in Chinese folk medicine for the treatment of gastritis (Duke and Ayensu, 1985). Al-Meshal *et al.* (1985) demonstrated that prophylactic treatment with fenugreek extract for 5 days did not produce any protective effect against gastric lesions induced by phenylbutazone and reserpine in rats. These results show the absence of any antisecretory and cytoprotective effect. However, when administered as a curative for five consecutive days to rats already treated with ulcerogenic doses of phenylbutazone, it produced a significantly faster healing of the ulcers. Fenugreek extract produced a mild relaxant effect on the smooth muscle of a rabbit's isolated duodenum when added to the organ-bath at 0.5 mg/mL (Al-Meshal *et al.*, 1985). The marked demulcent activity and mild anticholinergic action of fenugreek was suggested to be responsible for its effectiveness in promoting the healing of phenylbutazone induced ulcers.

Wound healing properties of fenugreek seeds have also been demonstrated in excision, incision and dead-space wound models in rats (Taranalli and Kuppast, 1996). Fenugreek seed suspension was more effective than aqueous seed extract in promoting wound healing in these models.

Anti-cancer effects

The ethanolic extract of *Trigonella foenum-graecum*, with an ED_{50} less than 10 μg/mL in the brine shrimp cytotoxicity assay, was also observed to possess anti-tumour activity in A-549 male lung carcinoma, MCF-7 female breast cancer and HT-29 colon adenocarcinoma cell lines (Alkofahi *et al.*, 1996). The extract gave negative results in the mutagenicity test.

Anti-microbial effects

Bhatti *et al.* (1996) reported that the aqueous and ethanol extracts of fenugreek seeds showed anti-bacterial activity.

Anthelmintic properties

Fenugreek seeds have been used as an anthelmintic against the most common nematodes (Mishra *et al.*, 1965). Ghafghazi *et al.* (1980) showed a water extract of fenugreek seeds to have dose dependent anthelmintic activity *in vitro* on both cestodes and nematodes. The extract also resulted in 87 per cent inhibition of embryonation of *Ascaris lumbricoides* eggs (Ghafghazi *et al.*, 1980).

Anti-nociceptive effects

Using the tail-flick and formalin tests, Javan *et al.* (1997) have demonstrated an anti-nociceptive effect of an aqueous extract prepared from fenugreek leaves (1–2 g/kg given intraperitoneally).

Toxicity studies

Short-term (90 days) feeding of fenugreek seeds to rats at levels equivalent to two and four times the therapeutic dose recommended for humans (25 g/day) produced no toxic effects as evidenced by: normal liver function tests, lack of any histopathological changes in the liver and no changes in haematological parameters (Udayasekhara Rao *et al.*, 1996). Moreover, long-term (24 weeks) administration of fenugreek seeds at 25 g/day, exhibited no clinical hepatic or renal toxicity or haematological abnormalities in diabetic subjects (Sharma *et al.*, 1996b). This dose was sufficient to improve glucose tolerance (Raghuram *et al.*, 1994; Sharma, 1986b) and lipid profile (Sharma *et al.*, 1996a) in NIDDM humans.

Two cases of severe reactions to fenugreek seed powder were reported in patients known to suffer from food allergies (Patil *et al.*, 1997). The first developed rhinorrhoea, wheezing and fainting following inhalation of the powder. The second developed numbness of the head, facial angioedema and wheezing after applying fenugreek paste to the scalp as a dandruff treatment. In skin scratch tests, a number of patients were found to have strong sensitivity to fenugreek. Immunoglobulins (IgE-type) capable of binding to proteins in fenugreek seeds were found in the sera of some patients.

Fenugreek seed extract did not produce any effect on the mean arterial blood pressure of anaesthetised rabbit in a dose of 20 mg i.v. nor on the isolated heart at a dose of 2.5 mg added to the perfusion fluid (Al-Meshal *et al.*, 1985).

The LD_{50} of fenugreek leaf aqueous extract in male and female mice was reported to be about 10 g/kg body weight for oral administration and 2 g/kg for intraperitoneal adminstration. Mild central nervous stimulation, rapid respiration and tremors were observed following high doses of the aqueous extract (Abdel-Barry *et al.*, 1997). Javan *et al.* (1997) estimate the LD_{50} in mice of a similar extract as 4 g/kg by the same route.

Summary

Trigonella foenum-graecum (fenugreek) is an important culinary and medicinal plant in many cultures. Fenugreek seeds have been widely studied for their reputed anti-diabetic, hypocholesterolaemic and anti-fertility effects. Various preparations of the seeds have been shown in human and animal model studies to lower blood glucose, improve glucose and starch tolerance and have beneficial effects on serum cholesterol and lipid profiles. The anti-diabetic effects have been associated with the intestinal effects of the gum fibre (galactomannan), insulin secretagogue activity of a major amino-acid (4-hydroxyisoleucine) and unidentified components with effects on peripheral glucose utilisation. Hypocholesterolaemic effects have been associated mainly with reduced intestinal reabsorption of cholesterol and bile acids. This activity has been linked to the saponins and sapogenins (e.g. diosgenin), and also to galactomannan fibre. However, hypolipidaemic effects are associated only with the saponins or sapogenins and not the fibre. Fenugreek steroidal sapogenins have been suggested to possess spermicidal and anti-androgenic activities in male rats, whilst crude fenugreek extracts have been reported to be abortifacient and cause embryo resorption in female rats.

Properties of fenugreek that have been reported but which have received less attention, include anti-cancer, anti-bacterial, anthelmintic, anti-cholinergic and ulcer and wound healing activities. Fenugreek leaves have been less well studied than the seeds, but are reported to have antinociceptive and hypoglycaemic effects.

The considerable body of scientific evidence reviewed here, suggests that fenugreek does indeed possess a number of important medicinal properties. The consumption of defatted fenugreek may be particularly beneficial in the management of diabetes and hypercholesterolaemia and the prevention of atherosclerosis and coronary heart disease. It may be advisable, however, to note the potential for anti-fertility effects and allergic reactions in susceptible individuals.

Acknowledgement

The authors thank the British Council (Sana'a, Yemen) for financing a sabbatical visit by Dr Molham Al-Habori to King's College London, during which time this manuscript was prepared. A review article containing much of the information presented in this chapter has been published in *Phytotherapy Research*. Al-Habori-M and Raman A (1998) Review: Antidiabetic and hypocholesterolaemic effects of fenugreek. *Phytotherapy Research* 12, 233–42.

References

Abdel-Barry, J.A., Abdel-Hassan, I.A. and Al-Hakiem, M.H.H. (1997) Hypoglycaemic and anti-hyperglycaemic effects of *Trigonella foenum-graecum* leaf in normal and alloxan induced diabetic rats. *J. Ethnopharmacol.* **58**, 149–55.

Ackerman, D.Z. (1912) Biol. 59, 17. *Per* Shani, J., Goldschmied, A., Ahronson, Z. and Sulman, F.G. (1974) Hypoglycaemic effect of *Trigonella foenum graecum* and *Lupinus termis* (Leguminosae) seeds and their major alkaloids in alloxan diabetic and normal rats. *Arch. Int. Pharmacodyn. Ther.* **210**, 27–36.

Ajabnoor, M.A. and Tilmisany, A.K. (1988) Effect of *Trigonella foenum graecum* on blood glucose levels in normal and alloxan-diabetic mice. *J. Ethnopharmacol.* **22**, 45–9.

Al-Hamood, M.H. and Al-Bayati, Z.F. (1995) Effect of *Trigonella foenum-graecum*, *Nerium oleander* and *Ricinus communis* on reproduction in mice. *Iraqi J. Sci.*, **36**, 436.

Al-Meshal, I.A., Parmar, N.S., Tariq, M. and Aqeel, A.M. (1985) Gastric anti-ulcer activity in rats of *Trigonella foenum graecum* (Hu-Lu-Pa). *Fitoterapia* **56**, 232–5.

Alarcon-Aguilar, F.J., Roman Ramos, R. and Flores Saenz, J.L. (1993) Plants medicinales usadas en el control de la diabetes mellitus. *Ciencia* **44**, 361–3.

Alcock, N.W., Crout, D., Gregorio, M., Lee, E., Pike, G. and Samuel, C.J. (1989) Stereochemistry of the 4-hydroxyisoleucine from *Trigonella foenum graecum*. *Phytochem.* **28**, 1835–41.

Ali, L., Azad Khan, A.K., Hassan, Z., Mosihuzzaman, M., Nahar, N., Nasreen, T., Nur-e-Alam, M. and Rokeya, B. (1995) Characterization of the hypoglycaemic effects of *Trigonella foenum graecum* seed. *Planta Med.* **61**, 358–60.

Alkofahi, A., Batshoun, R., Owais, W. and Najib, N. (1996) Biological activity of some Jordanian medicinal plant extracts. *Fitoterapia* **LXVII**, 435–42.

Amin, R., Abdul-Ghani, A.S. and Suleiman, M.S. (1987) Effect of *Trigonella foenum graecum* on intestinal absorption. *Diabetes* **36** (supp. 1), 211A.

Amin, R., Abdul-Ghani, A.S. and Suleiman, M.S. (1988) Effect of fenugreek and lupin seeds on the development of experimental diabetes in rats. *Planta Med.* **54**, 286–90.

Anis, M. and Aminuddin, E. (1985) Estimation of diosgenin in seeds of induced autoploid *Trigonella foenum graecum*. *Fitoterapia* **56**, 51–2.

Attaur-Rahman, A. and Zaman, K. (1989) Medicinal plants with hypoglycaemic activity. *J. Ethnopharmacol.* **26**, 1–55.

Baccou, J.C., Sauvaire, Y., Ollie, V. and Petit, L.J. (1978) L'huile de fenugreec, composition, properties, possibilities d'utilisation dans l'industrie des peintures et vernis. *Rev. Fr. des Corbs Gras.* **25**, 353.

Bailey, C.J. and Day, C. (1989) Traditional treatments for diabetes from Asia and the West Indies. Prac. *Diabetes* **3**, 190–2.

Berthoud, H.R., Bereiter, D.A., Trimble, E.R., Siegel, E.G. and Jeanrenaud, B. (1981) Cephalic phase, reflex insulin-secretion: neuroanatomical and physiological characterization. *Diabetolog.* **20**, 393–401.

Best, C.H. (1962) Epochs in the history of diabetes. In R.H. Williams (ed.) *'Diabetes'* Hueber, Ismaning, pp. 1–13.

Betteridge, D.J. (1989) Diabetes, lipoprotein metabolism and atherosclerosis. *Br. Med. Bull.* 45, 285–311.

Bever, B.O. and Zahnd, G.R. (1979) Plants with oral hypoglycaemic action. *Q. J. Crude Drug Res.* 17, 139–96.

Bhat, B.G., Sambaiah, K. and Chandrasekhara, N. (1985) The effect of feeding fenugreek and ginger on bile composition in the albino rat. *Nutr. Rep. Int.* 32, 1145–51.

Bhathena, S.J., Avigan, J. and Schreiner, M.E. (1974) Effect of insulin on sterol and fatty acid synthesis and HMG CoA reductase activity in mammalian cells grown in culture. *Proc. Natl. Acad. Sci.* 71, 2174–9.

Bhatti, M.A., Khan, M.T.J., Ahmed, B., Jamshaid, M. and Ahmad, W. (1996) Antibacterial activity of *Trigonella foenum-graecum* seeds. *Fitoterapia* **LXVII**, 372–4.

Birk, Y. (1969) Saponins. In I.E. Liener (ed.) *Toxic Constituents of Plant Foodstuffs*, Academic press, New York and London, pp. 169–210.

Blackburn, N.A., Redfern, J.S., Jarjis, H., Holgate, A.M., Hanning, I., Scarpello, J.H.B., Johnson, I.T. and Read, N.W. (1984) The mechanism of action of guar gum in improving glucose tolerance in man. *Clin. Sci.* 66, 329–36.

Brenac, P. and Sauvaire, Y. (1996) Accumulation of sterols and steriodal sapogenins in developing fenugreek pods: possible biosynthesis *in situ. Phytochem.* 41, 415.

Calvert, G.D., Blight, L., Illman, R.J., Topping, D.L. and Potter, J.D. (1981) A trial of the effects of soya-bean saponins on plasma lipids, faecal bile acids and neutral sterols in hypercholesterolaemic men. *Br. J. Nutr.* 45, 277–81.

Carlson, L.A. and Bottiger, L.E. (1985) Risk factors for ischaemic heart disease in men and women. *Acta. Med. Scand.* 18, 207–11.

Carlson, L.A., Bottiger, L.E. and Ahfeldt, P.E. (1979) Risk factors for myocardial infarction in the Stockholm prospective study. A 14-year follow up focusing on the role of plasma triglyceride and cholesterol. *Acta. Med. Scand.* 206, 351–60.

Cayen, M.N. and Dvornik, D. (1979) Effect of diosgenin on lipid metabolism in rats. *J. Lipid Res.* 20, 162–74.

Chatterjee, B.P., Sakar, N. and Rao, A.S. (1982) Serological and chemical investigations of the anomeric configuration of sugar units in the D-galacto-D-mannan of fenugreek (*Trigonella foenum graecum*) seeds. *Carbohyd. Res.* 104, 348–53.

Chinoy, N.J., Sheth, K.M. and Seethalakshmi, L. (1982) Studies on reproductive physiology of animals with special reference to fertility control. *Comp. Physiol. Ecol.* 7, 325–45.

Chopra, R.N., Chopra, I.C., Handa, K.L. and Kapur, L.D. (1982) *Chopra's Indigenous Drugs of India*, Academic Publishers, Calcutta, New Delhi, India, p. 582.

Desta, B. (1994) Ethiopian traditional herbal drugs. Part III: Anti-fertility activity of 70 medicinal plants. *J. Ethnopharmacol.* 44, 199–209.

Dhawan, B.N., Patnaik, G.K., Rastogi, R.P., Singh, K.K. and Tandon, J.S. (1977) Screening of Indian plants for biological activity. *Indian J. Exp. Biol.* 15, 208–19.

Dreher, M.L. (1987) *Handbook of Dietary Fibre. An Applied Approach*. Dekker, New York, USA, p. 199.

Duke, J.A. and Ayensu, E.A. (1985) *Medicinal Plants of China*. Reference Publications Inc., Michigan, USA, Vol. 1, p. 345.

Edwards, C.A., Johnson, I. and Read, N.W. (1988) Do viscous polysaccharides slow absorption by inhibiting diffusion or conversion? *Eur. J. Clin. Nutr.* 42, 307–12.

El-Mahdy, A.R. and El-Sebaiy, L.A. (1985) Proteolytic activity, amino acid composition, protein quality of fermented fenugreek seeds (*Trigonella foenum-graecum*). *Food Chem.* 18, 19–33.

El-Ridi, M.S., Azouz, W.M. and Hay, A.E. (1954) Isolation of a lactation promoting factor in fenugreek oil. *Hoppe-Seyler's J. Physiol. Chem.* 286, 256.

Elbetieha, A., Al-Hamood, M.H. and Alkofahi, A. (1996) Anti-implantation potential of some medicinal plants in female rats. *Arch. Std. Hiv. Res.* 10, 181–7.

Fazli, F.R.Y. and Hardman, R. (1968) The spice, fenugreek (*Trigonella foenum graecum*): its commercial varieties of seed as a source of diosgenin. *Trop. Sci.* 10, 66.

Fourier, F. (1948) *Plantes medicinales et venereuses de France. Paris* 111, 495.

Fowden, L., Pratt, H.M. and Smith, A. (1973) 4-Hydroxyisoleucine from seed of *Trigonella foenum graecum*. *Phytochem.* 12, 1701–7.

Gestetner, B., Assa, Y., Henis, Y., Birk, Y. and Bondi, A. (1971) Lucerne saponins IV: relationship between their chemical constituent and haemolytic and antifungal activities. *J. Sci. Food Agric.* 22, 168–72.

Ghafghazi, T., Farid, H. and Pourafkari, A. (1980) *In vitro* study of the anthelmintic action of *Trigonella foenum-graecum* grown in Iran. *Iranian J. Public Health* 9, 21–6.

Ghafghazi, T., Sheriat, H.S., Dastmalchi, T. and Barnett, R.C. (1977) Antagonism of cadmium and alloxan-induced hyperglycaemia in rats by *Trigonella foenum graecum*. *Shiraz. Med. J.* 8, 14–25.

Ghosal, S., Srivastava, R.S., Chatter, D.C. and Dutta, S.K. (1974) Extractives of *Trigonella*-1. Fenugreekine, a new stercal sapogenin-peptide ester of *Trigonella foenum graecum*. *Phytochem.* 13, 2247–51.

Gibney, M.J., Pathirana, C. and Smith, P. (1982) Saponins and fibre: lack of interactive effects on serum and liver cholesterol in rats and hamsters. *Atherosclerosis* 45, 365–7.

Girardon, P., Bessiere, J.M., Baccou, J.C. and Sauvaire, Y. (1985) Volatile constituents of fenugreek seeds. *Planta Med.* 6, 533–4.

Goda, T., Yamada, K., Sugiyama, M., Moriuchi, S. and Hoya, N. (1982) Effect of sucrose and acarbose feeding on the development of streptozotocin-induced diabetes in the rat. *J. Nutr. Sci. V.* 28, 41–56.

Gopalan, C., Rama Shastri, B.V. and Balasubramanyan, S.C. (1978) Nutritive value of Indian foods. National Institute of Nutrition, ICMR, Hyderabad, India.

Gordon, T., Castelli, W.P., Hjortland, M.E., Kannel, W.B. and Bawber, T.R. (1977) Predicting coronary heart disease in middle aged and older persons. *JAMA* 32, 497–504.

Grundy, S.M. (1987) Dietary treatment of hyperlipidaemia. In D. Steinberg and J.M. Olefsky (eds), *Hypercholesterolaemia and Atherosclerosis-Pathogenesis and Prevention*, Churchill Livingstone, London, p. 169.

Hemavathy, J. and Prabhakar, J.V. (1989) Lipid composition of fenugreek (*Trigonella foenum graecum* L.) seeds. *Food Chem.* 31, 1–7.

Hillaire-Buys, D., Petit, P., Manteghetti, M., Baissac, Y., Sauvaire, Y. and Ribes, G. (1993) A recently identified substance extracted from fenugreek seeds stimulates insulin secretion in rat. *Diabetolog.* 36, A119.

Holt, S., Heading, R., Carter, D.C., Prescott, L.F. and Tothill, P. (1979) Effect of gel fibre on gastric emptying and absorption of glucose and paracetamol. *Lancet* 20, 636–9.

Ivorra, M.D., Paya, M. and Villar, A. (1989) A review of natural products and plants as potential anti-diabetic drugs. *J. Ethnopharmacol.* 27, 243–75.

Javan, M., Ahmadiani, A., Semnanian, S. and Kamalinejad, M. (1997) Antinociceptive effects of *Trigonella foenum-graecum* leaves extract. *J. Ethnopharmacol.* 58, 125–9.

Jenkins, D.J.A. (1979) Dietary fibre, diabetes and hyperlipidaemia. *Lancet* 2, 1287–9.

Jenkins, D.J.A. and Jenkins, A. (1984) The clinical implication of dietary fibre. *Adv. Nutr. Res.* 6, 169–201.

Jenkins, D.J.A., Reynolds, D., Salvin, B., Leeds, A.R., Jenkins, A.L. and Jepson, E.H. (1980) Dietary fibre and blood lipids: treatment of hypercholesterolaemia with guar crispbread. *Am. J. Clin. Nutr.* 33, 575–81.

Jenkins, D.J.A., Wolever, T.M.S., Leeds, A.R., Gassull, M.A., Haisman, P., Dilawari, J, Goff, D.V., Metz, G.L. and Alberti, K.G.M.M. (1978) Dietary fibres, fibre analogues, and glucose tolerance: importance of viscosity. *Br. Med. J.* I, 1392–4.

Johnson, I.T. and Gee, J. (1980) Inhibitory effect of guar gum on the intestinal absorption of glucose *in vitro*. *Proc. Nutr. Soc.* 39, 52A.

Kamal, R., Yadav, R. and Sharma, J.D. (1993) Efficacy of the steroidal fraction of fenugreek seed extract on fertility of male albino rats. *Phytotherapy Res.* 7, 134–8.

Kannel, W.B. (1983) High-density lipoprotein: epidemiologic profile and risks of coronary artery disease. *Am. J. Cardiol.* 52, 9B.

Khare, A.K., Sharma, M.K. and Bhatnagar, V.M. (1983) Mild anti-fertility effect of ethereal extract of seeds of *Trigonella foenum-graecum* (Methi) in rats. *Arogya-J. Health Sci.* IX, 91–3.

Khosla, P., Gupta, D.D. and Nagpal, R.K. (1995a) Effect of *Trigonella foenum graecum* (fenugreek) on blood glucose in normal and diabetic rats. *Indian J. Physiol. Pharmacol.* 39, 173–4.

Khosla, P., Gupta, D.D. and Nagpal, R.K. (1995b) Effect of *Trigonella foenum graecum* (fenugreek) on serum lipids in normal and diabetic rats. *Indian J. Pharmacol.* 27, 89–93.

Kirby, R.W., Anderson, J.W. and Sieling, B. (1981) Oat bran intake selectively lowers serum low density lipoprotein cholesterol concentration of hypercholesterolaemic men. *Am. J. Clin. Nutr.* 34, 824–9.

Kohlrausch, A.Z. (1912) *Biol.* 57, 273. Per Shani, J., Goldschmied, A., Ahronson, Z. and Sulman, F.G. (1974) Hypoglycaemic effect of *Trigonella foenum graecum* and *Lupinus termis* (Leguminosae) seeds and their major alkaloids in alloxan diabetic and normal rats. *Arch. Int. Pharmacodyn. Ther.* 210, 27–36.

Kritchevsky, D. (1982) Fibre and lipids. In G.V. Vahouny and D. Kritchevsky (eds), *Dietary Fibre in Health and Disease*, Plenum Press, New York, p. 182.

Kritchevsky, D., Tepper, S.A. and Story, J. (1978) Influence of soya protein and casein on atherosclerosis in rabbits. *Fed. Proc.* 34, 747.

Laguna, J., Gomez-Puyou, A., Pena, A. and Guzman-Garcia, J. (1962) Effect of diosgenin on cholesterol metabolism. *J. Atherosclerosis Res.* 2, 459–70.

Lipid Research Clinics Program (1984) The lipid research clinics coronary primary prevention trial results. 1: Reduction in the incidence of coronary heart disease. *J. Am. Med. Assoc.* 251, 351–64.

Madar, Z. (1984) Fenugreek (*Trigonella foenum graecum*) as a means of reducing postprandial glucose level in diabetic rats. *Nutr. Rep. Int.* 29, 1267–73.

Madar, Z. and Odes, H.S. (1990) Dietary fibre in metabolic disease. In R. Paoletti (ed.) *Dietary Fibre Research*, Basel, Karger, pp. 1–54.

Madar, Z. and Shomer, I. (1990) Polysaccharide composition of a gel fraction derived from fenugreek and its effect on starch digestion and bile acid absorption in rats. *J. Agric. Food Chem.* 38, 1535–9.

Madar, Z., Abel, R., Samish, S. and Arad, J. (1988) Glucose lowering effect of fenugreek in non-insulin dependent diabetics. *Eur. J. Clin. Nutr.* 42, 51–4.

Mahato, S.B., Ganguly, A.N. and Sahu, N.P. (1982) Steroid saponins. *Phytochem.* 21, 959–78.

Maison, A.S. and Boucher, B.J. (1978) Diabetes mellitus. In R.B. Scott (ed.) *Price's Text Book of the Practice of Medicine ELBS*, Oxford Medical Publication, pp. 435–47.

Malinow, M.R. (1984) Triterpenoid saponins in mammals: effects on cholesterol metabolism and athero-sclerosis. In W.D. Nes, G. Fuller and L.S. Tsai (eds) *Biochemistry and Function of Isopentenoids in Plants*, Marcel Dekker, New York, pp. 229–46.

Malinow, M.R. (1985) Effects of synthetic glycosides on cholesterol absorption. *Ann. NY. Acad. Sci.* 454, 23–7.

Malinow, M.R., McLaughlin, P., Kohler, G.O. and Livingston, A.L. (1977) Prevention of elevated choles-terolaemia in monkeys by alfalfa saponins. *Steroids* 29, 105–10.

Malinow, M.R., McLaughlin, P. and Stafford, C. (1978) Prevention of hypercholesterolaemia in monkeys (*Macaca fascicularis*) by digitonin. *Am. J. Clin. Nutr.* 31, 814–18.

Malinow, M.R., Connor, W.E., McLaughlin, P., Stafford, C., Lin, D.S., Livingston, A.L., Kohler, G.O. and McNulty, W.P. (1981) Cholesterol and bile acid balance in *Macaca fascicularis*: effects of alfalfa saponins. *J. Clin. Invest.* 67, 156–62.

Malinow, M.R., Elliott, W.H., McLaughlin, P. and Upson, B. (1987) Effects of synthetic glycosides on steroid balance in *Macaca-Fascicularis*. *J. Lipid Res.* 28, 1–9.

Marles, R.J. and Farnsworth, N.R. (1995) Anti-diabetic plants and their active constituents. *Phytomedicine* 2, 137–89.

Max, B. (1992) This and that: the essential pharmacology of herbs and spices. *Trends Pharmacol. Sci.* 13, 15–20.

Miettinen, T.A. (1987) Dietary fibre and lipids. *Am. J. Clin. Nutr.* 45(suppl.), 1237–42.

Mishkinsky, J., Joseph, B. and Sulman, F. (1967) Hypoglycaemic effect of trigonelline. *Lancet* i, 1311–12.

Mishra, S.S., Tewari, J.P. and Saxena, K.B. (1965) Anthelmintic activity of some Indian medical plants. *Ind. J. Med. Sci.* 19, 398.

Mital, N. and Gopaldas, T. (1986a) Effect of fenugreek (*Trigonella foenum-graecum*) seed based diets on the birth outcome in albino rats. *Nutr. Rep. Int.* 33, 363–9.

Mital, N. and Gopaldas, T. (1986b). Effect of fenugreek (*Trigonella foenum-graecum*) seed based diets on the lactational performance in albino rats. *Nutr. Rep. Int.* **33**, 477–84.

Moissides, M. (1939) Le fenugrec autrefois et aujourd'hui. *Janus* **43**, 123–30.

Monnier, L.H., Pham, T.C., Aguirre, L., Orsetti, A. and Mirouze, J. (1978) Influence of indigestible fibres on glucose tolerance. *Diabetes Care* **1**, 83–8.

Moorthy, R., Prabhu, K.M. and Murthu, P.S. (1989) Studies on the isolation and effect of an orally active hypoglycaemic principle from the seeds of fenugreek (*Trigonella foenum graecum*). *Diabetes Bull.* **9**, 69–72.

Nadkarni, K.M. (1954) *Indian Materia Medica*, Popular book depot, Bombay. Vol. I, pp. 1240–3.

National Institute of Nutrition (1987) Annual Report, Indian Council of Medical Research, Hyderabad, India, p. 11.

O'Connor, N., Tredger, J. and Morgan, L. (1981) Viscosity differences between various guar gums. *Diabetolog.* **20**, 612–15.

Oakenful, D.G. and Fenwick, D.E. (1978) Adsorption of bile salts from aqueous solution by plant fibre and cholestyramine. *Br. J. Nutr.* **40**, 299–309.

Oakenful, D.G., Topping, D.L., Illman, R.J. and Fenwick, D.E. (1984) Prevention of dietary hypercholesterolaemia in the rat by soybean and quillaya saponins. *Nutr. Rep. Int.* **25**, 1039–46.

Paisey, R.B., Arredondo, G., Villalobos, A., Lozano, O., Guevara, L. and Kelly, S. (1984) Association of differing dietary, metabolic and clinical risk factors with macrovascular complications of diabetes. A prevalence study of 503 Mexican type II diabetic subjects 1. *Diabetes Care* **7**, 421–7.

Parry, J.W. (1943) *The Spice Handbook*, Chemical Publishing Co., Brooklyn, New York.

Patil, S.P., Niphadkar, P.V. and Bapat, M.M. (1997) Allergy to fenugreek (*Trigonella foenum graecum*). *Ann. Allergy, Asthma and Immunol.* **78**(3), 297–300.

Petit, P., Sauvaire, Y., Ponsin, G., Manteghetti, M., Fave, A. and Ribes, G. (1993) Effect of a fenugreek seed extract on feeding behaviour in the rat: metabolic-endocrine correlates. *Pharmacol. Biochem. Behav.* **45**, 369–74.

Petit, P., Sauvaire, Y., Hillaire-buys, D., Manteghetti, M., Baissac, Y., Gross, R. and Ribes, G. (1995a) Insulin stimulating effect of an original amino acid, 4-hydroxyisoleucine, purified from fenugreek seeds. *Diabetolog.* **38**(S1), A101.

Petit, P., Sauvaire, Y., Hillaire-buys, D., Leconte, O., Baissac, Y., Ponsin, G. and Ribes, G. (1995b) Steriod saponins from fenugreek seed: extraction, purification and pharmacological investigation on feeding behaviour and plasma cholesterol. *Steroids* **60**, 674–80.

Platel, K. and Srinivasan, K. (1996) Influence of dietary spices or their active principles on digestive enzymes of small intestinal mucosa in rats. *Int. J. Food Sci. Nutr.* **47**, 55–9.

Prasad, M.R. and Rajalakshmi, M. (1976) Target sites of suppressing fertility in the male. In R.L. Singer and J.A. Thomas (eds) *Cellular Mechanism Modulating Gonadal Action*, University Park Press, Baltimore. Vol. 2, p. 263.

Price, K.R., Johnson, I.T. and Fenwick, G.R. (1987) The chemistry and biological significance of saponins in foods and feedingstuffs. *CRC Crit. Rev. Food Sci. Nutr.* **26**, 27–135.

Raghuram, T.C., Sharma, R.D., Sivakumar, B. and Sahay, B.K. (1994) Effect of fenugreek seeds and intra-venous glucose disposition in non-insulin dependent diabetic patients. *Phytother. Res.* **8**, 83–6.

Rao, V.S., Menezes, A.M. and Gadelha, M.G. (1988) Anti-fertility screening of some indigenous plants of Brazil. *Fitoterapia* **LXI**, 17–20.

Reid, J.S.G. and Meier, H. (1970) Formation of reserve galactomannan in the seeds of *Trigonella foenum graecum. Phytochem.* **9**, 513–20.

Report of the National Cholesterol Education Program (1988) Expert panel on detection, evaluation and treatment of high blood cholesterol in adults. *Arch. Int. Med.* **148**, 36–69.

Ribes, G., Da Costa, C., Loubatieres-Mariani, M.M., Sauvaire, Y. and Baccou, J.C. (1987) Hypocholesterolaemic and hypotriglyceridaemic effects of subfractions from fenugreek seeds in alloxan diabetic dogs. *Phytother. Res.* **1**, 38–43.

Ribes, G., Sauvaire, Y., Baccou, J.C., Valette, G., Chenon, D., Trimble, E.R. and Loubatieres-Mariani, M.M. (1984) Effects of fenugreek seeds on endocrine pancreatic secretions in dogs. *Ann. Nutr. Metab.* **28**, 37–43.

Ribes, G., Sauvaire, Y., Da Costa, C., Baccou, J.C. and Loubatieres-Mariani, M.M. (1986) Antidiabetic effects of subfractions from fenugreek seeds in diabetic dogs. *Proc. Soc. Exp. Biol. Med.* **182**, 159–66.

Sadhukhan, B., Roychowdhury, U., Banerjee, P. and Sen, S. (1994) Clinical evaluation of a herbal anti-diabetic product. *J. Indian Med. Assoc.* **92**, 115–17.

Sankara Rao, D.S. and Deosthale, Y.G. (1981) Mineral composition of four Indian food legumes. *J. Food Sci.* **46**, 1962–3.

Sauvaire, Y. and Baccou, J.S. (1976) Nutritional value of the proteins of leguminous seed, fenugreek (*Trigonella foenum graecum* L.). *Nutr. Rep. Int.* **14**, 527–35.

Sauvaire, Y., Baissac, Y., Leconte, O., Petit, P. and Ribes, G. (1996) Steroid saponins from fenugreek and some of their biological properties. *Adv. Exp. Med. Biol.* **405**, 37–46.

Sauvaire, Y., Girardon, P., Baccou, J.C. and Risterucci, A.M. (1984) Changes in growth, proteins and free amino acids of developing seed and pod of fenugreek. *Phytochem.* **23**, 479–86.

Sauvaire, Y., Ribes, G., Baccou, J.C. and Loubatiers-Mariani, M.M. (1991) Implication of steroid saponins and sapogenins in the hypocholesterolaemic effect of fenugreek. *Lipids* **26**, 191–7.

Sethi, N., Nath, D., Singh, R.K. and Srivastava, R.K. (1990) Anti-fertility and teratogenic activity of some indigenous medicinal plants in rats. *Fitoterapia* **LXI**, 64–7.

Setty, B.S., Kamboj, V.P., Garg, H.S. and Khanna, N.M. (1976) Spermicidal potential of saponins isolated from Indian medicinal plants. *Contraception* **14**, 571–8.

Shani, J., Goldschmied, A., Ahronson, Z. and Sulman, F.G. (1974) Hypoglycaemic effect of *Trigonella foenum graecum* and *Lupinus termis* (Leguminosae) seeds and their major alkaloids in alloxan diabetic and normal rats. *Arch. Int. Pharmacodyn. Ther.* **210**, 27–36.

Shankaracharya, N.B. and Natarajan, C.P. (1972) Fenugreek: chemical composition and uses. *Indian Species*, IX(1), *op. cit.* Mital and Gopaldas (1986a).

Sharma, R.D. (1984) Hypocholesterolaemic activity of fenugreek (*T. foenum graecum*): an experimental study in rats. *Nutr. Rep. Int.* **30**, 221–31.

Sharma, R.D. (1986a) An evaluation of hypocholesterolaemic factor of fenugreek seeds (*T. foenum graecum*) in rats. *Nutr. Rep. Int.* **33**, 669–77.

Sharma, R.D. (1986b) Effect of fenugreek seeds and leaves on blood glucose and serum insulin responses in human subjects. *Nutr. Res.* **6**, 1353–64.

Sharma, R.D. and Raghuram, T.C. (1990) Hypoglycaemic effect of fenugreek seeds in non-insulin dependent diabetic subjects. *Nutr. Res.* **10**, 731–9.

Sharma, R.D., Raghuram, T.C. and Rao, N.S. (1990) Effect of fenugreek seeds on blood glucose and serum lipids in type I diabetes. *Eur. J. Clin. Nutr.* **44**, 301–6.

Sharma, R.D., Raghuram, T.C. and Rao, V.D. (1991) Hypolipidaemic effect of fenugreek seeds: a clinical study. *Phytother. Res.* **5**, 145–7.

Sharma, R.D., Sarkar, A., Hazar, D.K., Misra, B., Singh, J.B., Maheshwari, B.B. and Sharma, S.K. (1996a) Hypolipidaemic effect of fenugreek seeds: a chronic study in non-insulin dependent diabetic patients. *Phytother. Res.* **10**, 332–4.

Sharma, R.D., Sarkar, A., Hazar, D.K., Misra, B., Singh, J.B. and Maheshwari, B.B (1996b) Toxicological evaluation of fenugreek seeds: a long term feeding experiment in diabetic patients. *Phytother. Res.* **10**, 519–20.

Sidhu, G.S. and Oakenful, D.G. (1986) A mechanism for the hypocholesterolaemic activity of saponins. *Br. J. Nutr.* **55**, 643–9.

Sidhu, G.S., Upson, B. and Malinow, M.R. (1987) Effects of soy saponins and tigogenin cellobioside on intestinal uptake of cholesterol, cholate and glucose. *Nutr. Rep. Int.* **35**, 615–23.

Singhal, P.C., Gupta, R.K. and Joshi, L.D. (1982) Hypocholesterolaemic effect of *Trigonella foenum graecum* (Methi). *Current Sci.* **51**, 136–7.

Sparks, J.D. and Sparks, C.E. (1994) Insulin regulation of triacylglycerol-rich lipoprotein synthesis and secretion. *Biochim. Biophys. Acta* **1215**, 9–32.

Stark, A. and Madar, Z. (1993) The effect of an ethanol extract derived from fenugreek (*Trigonella foenum graecum* L.) on bile acid absorption and cholesterol levels in rats. *Br. J. Nutr.* **69**, 277–87.

Stolzenberg, S.J. and Parkhurst, R.M. (1974) Spermicidal actions of extracts and compounds from *Phytolacca dodecandra*. *Contraception* **10**, 135–43.

Story, J.A., Le Pages, S.L., Petro, M.S., West, L.G., Cassidy, M.M., Lightfoot, F.G. and Vahouny, G.V. (1984) Interactions of alfalfa plant and sprout saponins with cholesterol *in vitro* and in cholesterol-fed rats. *Am. J. Clin. Nutr.* **39**, 917–29.

Taranalli, A.D. and Kuppast, I.J. (1996) Study of wound healing activity of seeds of *Trigonella foenum grae-cum* in rats. *Ind. J. Pharm. Sci.* **58**(3), 117–19.

Trowell, H. (1972) Ischaemic heart disease and dietary fibre. *Am. J. Clin. Nutr.* **25**, 926–31.

Uchida, K., Takase, H., Nomura, Y., Takeda, K., Takeuchi, N. and Ischikawa, Y. (1984) Changes in biliary and faecal bile acids in mice after treatment with disogenin and B-sitosterol. *J. Lipid Res.* **25**, 236–45.

Udayasekhara Rao, P. and Sharma, R.D. (1987) An evaluation of protein quality of fenugreek seeds (*Trigonella foenum graecum*) and their supplementary effects. *Food Chem.* **24**, 1–9.

Udayasekhara Rao, P., Sesikeran, B. and Srinivasa Rao, P. (1996) Short term nutritional and safety evaluation of fenugreek. *Nutr. Res.* **16**, 1495–505.

Ulloa, N. and Nervi, F. (1985) Mechanism and kinetic characteristics of the uncoupling by plant steroids of biliary cholesterol from bile salt output. *Biochim. Biophys. Acta.* **837**, 181–9.

Valette, G., Sauvaire, Y., Baccou, J.C. and Ribes, G. (1984) Hypocholesterolaemic effect of fenugreek seeds in dogs. *Atherosclerosis* **50**, 105–11.

Varshney, I.P. and Sharma, S.C. (1996) Saponins XXXII *Trigonella foenum graecum* seeds. *J. Indian Chem. Soc.* **43**, 564–7.

Vinik, A.I. and Jenkins, D.J.A. (1988) Dietary fibre in management of diabetes. *Diabetes Care* **11**, 160–73.

West, K.M. (1978) *Epidemology of Diabetes and Its Vascular Complications*, Elsevier, New York.

West, K.M., Ahuja, M.M.S. and Bennet, P.H. (1983) The role of circulating glucose and triglyceride concentrations and their interactions with other 'risk factors' as determinants of arterial disease in nine diabetic population samples from the W.H.O. multinational study. *Diabetes Care* **6**, 361–9.

Wong, S., Traianedes, K. and O'Dea , K. (1985) Factors affecting the rate of hydrolysis of starch in legumes. *Am. J. Clin. Nutr.* **42**, 38–43.

Yoshikawa, M., Murakami, T., Komatsu, H., Murakami, N., Yamahara, J., Matsuda, H. (1997) Medicinal foodstuffs. IV. Fenugreek seed. (1): structures of trigoneosides Ia, Ib, IIa, IIb, IIIa and IIIb, new furostanol saponins from the seeds of Indian *Trigonella foenum-graecum* L. *Chem. Pharm. Bull.* **45**(1), 81–7.

11 Marketing

Christos V. Fotopoulos

Introduction

From time immemorial, spices have played a vital role in world trade due to their varied properties and applications. We primarily depend on spices for flavor and fragrance as well as for color, as a preservative and for its inherent medicinal qualities. Although about 107 spices are recorded, only about a dozen are important – black pepper, cardamom, ginger, turmeric, large cardamom, cumin, coriander, fennel, fenugreek, chillies, saffron and celery. Of all those spices the marketing analysis here will focus on fenugreek, although problems frequently arise with production and trade statistics since spice products are frequently combined under one heading (Edison, 1995).

Although the spice industry has undergone substantial changes since early developments, the product range and the global pattern of trade has not altered radically.

At the beginning of the twentieth century, Asian producers had achieved a dominant position in the export of spices, British India was by far the most important of these followed by Japan, Thailand, China and Dutch East Indies (now Indonesia). The main flow of trade was to Ceylon (Sri Lanka), which was the hub of the Asian market, and to the British Straits Settlement (now Malaysia) in which Singapore played an important role as an entrepot. Asian exports to Europe and North America were on a much smaller scale.

Severe disruption of the South East Asian and Far Eastern trade occurred during the Second World War. After the cessation of hostilities a rapid recover occurred and in the postwar period the main flow of trade in spices has been from India and China to Sri Lanka and Malaysia, and from Mexico and Japan to the US. From the early 1970s however, historical trading patterns underwent a significant change with the reduction of imports into Sri Lanka and the emergence of China and India as the world's chief exporters of spices, while Morocco is the second most important exporter to the European Union (EU) (Purseglove, 1981).

Although, historically, the spice industry in each of the main European nations developed to a large extent independently, the creation of the EU has done much to encourage its integration. Rotterdam, Hamburg, London and Marseilles have traditionally been the main entrepot centers for spices and many of the biggest importers are based in these cities. Some of these traders have themselves diversified into the processing and packing of spices. The majority of these companies are involved in importing other commodities and food stuffs. Some, however, specialize almost exclusively in one or two particular spices. All of them now operate on a European-wide basis.

The volume of world trade in fenugreek has always been subject to considerable fluctuations. One major factor contributing to these variations is that international trade in this commodity is only a small percentage of global production. Considerable difficulties are encountered in attempting to determine the level of trade in fenugreek. Apart from the common shortcomings in the export statistics of many of the major exporting countries, the trade in small volumes of this commodity from numerous other minor exporters is rarely reported, and even in some of the major importing countries in the Western hemisphere import statistics are frequently deficient. The published statistics must be regarded therefore, as no more than very approximate orders of magnitude in many instances. An estimate of the world trade in whole and ground fenugreek has fluctuated around 10,000 tons.

The overall market structure for fenugreek in Western Europe and the US are not dissimilar although there are differences at the margin. Common to other spices, both markets show some decline in the importance of brokers and agents as increasing direct contact is made by importers and spice-packers with suppliers in the producing countries. The two markets also show a decline in forward contracting in favor of spot trading.

Production and processing

Fenugreek is one of the earliest spices known to man. Ancient Egyptians used it as a food, medicine and embalming agent.

Fenugreek belongs to the legume family; it is a cripping plant (in some cases) with whitish blossoms. It is especially resistant to drought and temperature changes. As in the raze, with legumes, the whole plant and particularly the product and the seeds are rich in proteins. Fenugreek has a strong, pleasant and quite peculiar odor reminiscent of maple.

It is an annual, maturing about 3–5 months from sowing. During the annual production, the whole plant is harvested and hung up to dry before being threshed to obtain the square-shaped seed. International dealers require low levels of admixture (loose husks, dirt, other seeds, etc). The level should be no more than 4 percent and preferably below 1 percent (Robbins, 1997).

Use

In many infertile areas, the plant has been used (especially in early times) as an alternative to cereals in rotation techniques. Fenugreek fixes nitrogen in the soil and can be used as a forage as well as for the provision of seed. Forage yields of 9 tons/h and seed yields of 3.5 tons/h are claimed.

The main international trade in fenugreek is in the seeds but the fresh and dried leaves are also used to flavor curries.

The principal uses of fenugreek seed are in spice mixes for processed meat products and to a lesser extent, in curry powder. Fenugreek seed is also used extensively in Italian cooking, particularly in pizzas and certain pastas. The whole seed is available in retail packs. Other uses of fenugreek seed is in animal feed flavor for both ruminant and pig feed. Before incorporation in the feed, the seed is ground and roasted. Fenugreek was traditionally blended in equal proportions with aniseed but the price of aniseed has increased considerably and its use has been much reduced. Cheaper synthetics, including vanillin and anethole, have made inroads at the expense of natural spices, but fenugreek seed being reasonably low-priced has been able to maintain its position better than most (Smith, 1982).

An essential quantity of fenugreek seed is used for the production of extracts. Fenugreek spice extracts were developed to meet the new demands of the food processing industry. They have the following advantages:

- consistency in flavor
- not affected by bacterial contamination
- much longer shelf life

- easier storage and handling
- full release of flavor during cooking
- can easily be blended to achieve the desired characteristics.

The essential constituent of spices, which provides the aroma, flavor, pungency and colour, together make up a very small part, often less than 10 percent by weight of the whole. The balance mainly functions as the inert matrix and protective sheath for these essential constituents. These essential constituents may be obtained by solvent extraction of the spices, resulting in an extract called the spice oil or oleoresin, which consists of a complex mixture closely resembling the characteristics of the spice as a whole.

On steam distillation, the spices yield their volatile constituents. The essential oils thus obtained are endowed with the major part of the spice flavor and fragrance properties.

The oleoresins containing all the volatile as well as non-volatile constituents of the spices, most closely represent the total flavor of the fresh spice in a highly concentrated form (Spices Board India, 1997).

Fenugreek oil and oleoresin can be used to advantage wherever fenugreek spice is used, except in those applications where the appearance or filler aspect of the fenugreek spices is of importance. In addition to the benefit of standardization, consistency and hygiene afford by fenugreek oil and oleoresin, there is a big potential in their use for new product development. New flavors and fragrances are constantly being sought to entice the consumer. This applies equally to food products, medications, as well as other non-food products.

Fenugreek oil and oleoresin are mainly used as food flavors, especially in dressings, soups, packed goods, fish and vegetables. It can also be used in artificial maple syrups, cosmetics, in tobacco flavors and sour spice seasonings. Small quantities, with a declining trend of fenugreek extract, are used in animal feed flavors. The decline is attributed to competition from cheaper synthetics. The extract was usually blended with anethole or an aniseed extract and dispersed on a base for mixing with the feed. There have recently been technical developments involving the spraying of liquid flavors on the feed stuffs, which it is claimed gives a better flavor dispersion than the usual method of simply sprinkling the dry flavor compound on to the feed. Therefore the demand for fenugreek extract may increase again. However, there is still some resistance to liquid flavors for the reasons mentioned before. Furthermore, any increase in the use of fenugreek extract can be expected to lead to a corresponding fall in the use of the ground spice. Moreover, it is argued that the seed offers a great potential as a source of the steroid precursor diosgenin. However, despite the development of seeds with high diosgenin content, extraction is not yet economic.

Industry structure

Apart from the large trading houses there are a series of small importers (often of ethnic origin) who supply either whole or ground spices to health food shops, small grocers and market traders. As health and sanitary legislation becomes more rigorous it will be more and more difficult for these small companies to survive. They are presently the targets of much criticism concerning quality control and product testing methods.

Most spice grinders and packers in Europe were originally established as small family concerns. Many of them have now been sold to large, often multinational companies, specializing in spices and other food ingredients. The consolidation of the industry is taking place very rapidly. Small companies can no longer afford the very high capital costs of new processing and packing machinery and above all sophisticated testing and quality control equipment. Probably of greater importance is the growing cost of marketing and promotion. Only the larger food manufacturers can afford the enormous advertising and promotion costs involved in selling branded products. The market is increasingly dominated by two food groups: McCormick of the USA and Burns Philip and Co. of Australia. Other major companies include Fuchs (which operates in Germany and France), Ducros (which operates in France and Spain) and CPC International, a US company. Many of the smaller companies prefer to supply to the catering trade or pack on contract for the supermarkets.

The spice extraction industry, producing spice oils, oleoresins or concentrated spice extracts and flavors, is now mainly in the hands of companies manufacturing a range of food ingredients or flavors and fragrance compounds. Food ingredient manufacturers will produce such products as colorants, stabilizers, gum resins and emulsifiers as well as spice extracts. Many of these are still small independent firms (e.g. East Anglia Food Ingredients, UK, or Aralco, France). The industry reports a slow trend away from processed spice extracts to the natural product. People prefer to see the spices they are consuming in processed foods rather than taste invisible flavors.

Almost all the flavors and fragrance companies now operate on a global scale, producing customized flavor compounds for the large food manufacturers. Ten companies have more than 70 percent of this market worldwide. They include Quest Harmman and Reimer, Givaduan and International Flavors and Fragrances (Commonwealth Secretariat, 1996).

The rapid growth in convenience foods and the spread of fast-food chains will have a powerful influence on the future structure and direction of the spice industry. The ready-to-eat food and catering sector are in many cases larger consumers of spices and spice products than the household market. Many of the spice processors are themselves diversifying into food processing and food ingredient manufacturing. Companies like McCormick, Kuhne and Amora all supply pickles, relishes and mayonnaise as well as a wide range of pourable spice sauces. It is in this area, not in packaged spices that most observers see growth in the market.

The structure of the spice industry is presented analytically in Figure 11.1.

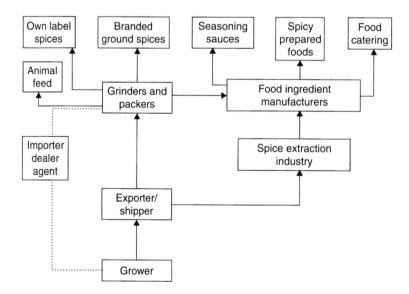

Figure 11.1 The structure of the spice industry.

Source: Commonwealth Secretariat, 1996.

Market structure of the main exporting and importing countries

Fenugreek is traded mainly in seed form and to a lesser extent as a spice and as an extract (oil, oleoresin). However, all three forms of traded fenugreek are often aggregated with other seeds, spices or extracts in trade statistics thus impeding the exact calculation of fenugreek traded volume. Here, an attempt is made to outline the market structure for fenugreek products in the major importing and exporting countries.

Exporting countries

India

India has a predominant position in the world spice trade with substantial production back up and availability of a wide range of spices. India produces over two million tons of spices every year. The total world trade in spices is only one-fifth of India's spice production. India is the largest supplier accounting for more than one-third of the total world spice trade of 450,000 tons. Indian spices are exported to over 130 countries. India is a major supplier of a large number of seed spices such as coriander, cumin, celery, fennel, fenugreek, garlic, etc. India is also the leading manufacturer and supplier of spice oil and oleoresins (Spices Board, 1996a).

Spice exports from India until recently were in raw form and in bulk packaging. The recent changes in market behavior, changes in consumer preferences and the emergence of super-markets, etc. abroad have resulted in the usage of more value added, ready to use spices products and spices in consumer packs. The main technology capabilities that India could achieve in the field of spice processing and post harvest handling have helped it to move ahead of other producing countries. The Indian exports in value added forms have shown significant growth during the years of the last decade. The exports of value added spices like spice oils and oleoresins, spice powders and mixtures, dehydrated spice products, etc., including spices in branded consumer pack, have substantially increased.

As shown in Table 11.1, the cultivated acreage of fenugreek and the respective production exhibit relative stability in the last twenty years, variation is small in acreage (25,000–30,000 ha have been cultivated) and slightly larger in production (35,000–45,000 tons have been produced) depending on weather conditions. However, exports exhibit an increasing trend, rather dramatic in recent years; stalling at 799 tons in 1960–61, exports rose to 15,135 tons in 1995–96, while export prices (in Rs/kgr) rose fifteen-fold during the same period. This increase in the quantity and value of fenugreek exports, in recent years, reflects improvements in the processed fenugreek products as well as production of new, high value-added ones.

Table 11.2 presents the major countries to which Indian fenugreek products are exported; most exports are directed to UAE, Sri Lanka and Japan. Of the EU countries the UK, the Netherlands, Germany and France are the major importing countries of Indian fenugreek products.

The Spices Board India (Ministry of Commerce) Government of India is the apex agency for the development and worldwide promotion of Indian spices. The Board is the catalyst of these dramatic transitions. The Board has been with the Indian Spice Industry every step of the way. The Board plays a far-reaching and influential role as a developmental, regulatory and promotional agency for Indian spices.

The Board is an international link between the Indian exporters and the importers abroad. Its broad-based activities include formulation and implementation of better production and quality improvement methods, systematic research and development programs, education and training

Table 11.1 Area, production and export of fenugreek from India

Year	Area (ha)	Production (tons)	Export (tons)	Export value (Rs/kgr)
1960–1			799	0.87
1970–1			1,042	1.40
1975–6	31,164	43,473	1,541	2.58
1976–7	32,964	49,659	1,873	2.36
1977–8	54,764	56,773	3,728	3.38
1978–9	31,276	48,176	5,256	3.59
1979–80	41,797	57,575	4,798	3.26
1980–1	38,478	52,636	4,470	3.80
1981–2	32,355	63,203	3,242	4.13
1982–3	32,246	45,697	3,967	4.24
1983–4	40,630	36,429	3,967	4.24
1984–5	44,687	53,580	5,545	4.95
1985–6	30,256	31,953	2,394	4.13
1986–7	23,866	25,949	3,224	5.22
1987–8	24,091	21,243	2,194	9.11
1988–9	38,402	37,431	3,575	10.26
1989–90	37,635	38,806	6,020	7.09
1990–1	37,297	37,694	3,748	8.13
1991–2	26,050	25,485	6,375	8.74
1992–3	24,629	25,372	5,255	10.84
1993–4	29,578	30,432	4,934	14.62
1994–5	38,633	49,046	7,956	15.40
1995–6			15,135	12.38

Source: Spices Board, 1996b.

Table 11.2 Fenugreek spice exports from India during 1991–2 to 1995–6 (QTV in MT, it is referred to main countries)

Countries	1991–2	1992–3	1993–4	1994–5	1995–6
Canada	32.7	16.6	37.0	102.6	111.8
France	8.0	47.7	145.0	172.0	242.0
Germany	53.5	117.1	155.0	182.2	203.2
Israel	102.5	125.3	163.3	282.5	338.3
Japan	853.2	425.0	780.4	1,065.8	401.5
Jordan	103.0	125.0	23.0	5.0	224.0
Korea (South)	168.0	277.9	164.5	230.0	250.0
Malaysia	241.7	169.9	96.3	191.1	305.3
The Netherlands	146.3	319.1	275.7	462.8	552.2
Singapore	992.9	437.0	479.8	415.7	418.5
Sri Lanka	664.0	102.0	474.0	1,204.7	1,237.6
Saudi Arabia	591.2	385.4	338.5	487.3	574.5
USA	457.3	461.8	219.8	462.4	668.3
UK	320.5	238.5	542.5	335.5	593.3
UAE	842.5	1,593.2	599.6	1,058.5	2,770.6
Total[*]	5,577.3	48,415.0	4,494.4	6,678.1	8,891.1

Source: Spices Board, 1996b.

Note

[*] Figures of exports only partly agree with the respective figures of Table 11.1 because only the major exporting destinations are included here.

of growers, processors, packers and exporters, selective registration and licensing. It acts as a data bank and communication channel for importers and exporters and promotes Indian spices abroad (Spices Board, 1996c).

The global food market is flush with all manner of branded spices in consumer packs. All of them bombard the consumer with chains and counterclaims for visibility and attention. But then, the packs seldom reveal the source of origin of the products nor do they offer a clue as to the quality associated with it. The result is that the consumer is totally confused. The Indian Spices Logo[1] is a major effort to overcome this impasse. The international consumer is by and large aware of the intrinsic and acquired superiority of Indian spices. The Board awards the logo selectively to exporters who have certified processing and quality control capability and maintain a high level of hygiene and sanitation at all stages.

The latest in the Board's campaign for quality upgradation is the introduction of the spice House Certificate. It is an effort to recognize those exporters who have a commitment to quality, consistency and long-term export growth. The certificate is issued to those processors/exporters who have adequate capabilities for cleaning, processing, grading, packaging, warehousing and quality assurance. It is hoped and believed that these units will move towards HACCP and ISO 9000 (Fotopoulos, 1995; Spices Board, 1996c).

The Spice Board has published two lists of exporting companies specializing in fenugreek seed and fenugreek powder, respectively. The first list consists of twenty-five companies including: Hathibhai Bulakhidas, Jatin and Company, Groversons, Swani Corporation, Gautam Export Corporation, Palbro International, etc. The second list includes fifteen companies such as: Vallabhadas Kanji Ltd, Vasantham Enterprises, Allana Sons Ltd, Shashuat Gum Industries, Miltop Exports, etc.

Morocco

Full statistics of fenugreek seed are only available from 1976–78 but these show annual exports varying between 700 and 1,700 tons. From an inspection of the statistics of the importing countries it appears likely that annual Moroccan exports have been around 1,000 tons. The main market has usually been Italy although Moroccan and Italian trade figures do not correspond. The UK has also been an important market. Other significant markets include the Netherlands, France, Germany, USA and Libya. Morocco exports small quantities of fenugreek extract mainly to France (Smith, 1982).

Other exporting countries

Many other countries export fenugreek seed from time to time but not in volumes comparable with those of India and Morocco. Spain has been a major supplier to the important Italian market, in some years supplying 100–200 tons. Tunisia, Turkey and Lebanon have also exported sizable quantities, but intermittently. In Asia, where the crop is widely produced for domestic usage, China and Pakistan, among others, have exported fenugreek seed but the quantities are much less than those for India. Elsewhere, Israel and Egypt occasionally export small quantities. Cultivation trials have been conducted in several countries including Ethiopia, Kenya, Tanzania

---

1 The logo, a green leaf inside an elliptical ring (denoting freshness, growth and excellence), is prominently displayed on all packs cleared and approved by the Spices Board India, so it can be easily spotted that the pack spells Indian and quality.

and even the UK, but as yet there has emerged no producer large enough to challenge the dominance of the two principal exporters in international markets.

Importing countries

Germany

Germany is the largest market for spices in Europe. There are more than sixty companies that are involved in the grading, packing and processing of spices, with another fifteen or more importers and distributors. Germany is also the largest importer, accounting for around 30 percent of the ECU 320 million European imports market. Fenugreek imports are not separated from turmeric imports trade statistics but from an examination of origins and of the export statistic for the source countries, it appears that perhaps about 200 tons are imported annually. Most of this is supplied by India but smaller amounts come from Morocco and China. The main use is in curry powders and other spice mixes but most of the consumption is probably accounted for in animal feed flavors. A small quantity of fenugreek is said to be imported annually from France for special application in tobacco flavors (Spices Board, 1996b; Commonwealth Secretariat, 1996).

German households have the highest per capita consumption of spices in Europe. Spiced bread and bakery products are widely consumed. The Germans are also the largest producers and consumers of processed meat products. These require a wide range of spices for both flavoring and coloring purposes. It can be pointed out here that Germany is also a major exporter of spices and spice products.

There is a growing concentration of retailing in the hand of the supermarkets, with European giants like Tengelman, Metro and Rewe becoming increasingly important (ten companies account for 70 percent of the turnover).

In addition, discount stores like Aldi operate throughout Europe. A similar concentration has taken place in the food processing and catering sector. This has given rise to a corresponding rationalization process amongst the producers and processors of spice (Commonwealth Secretariat, 1996).

France

France is the second largest spice market in Europe with a representation of 13 percent of the total EU market. France has over 15 percent of the EU import market, second only to Germany. French trade statistics aggregate imports with those of turmeric but an examination of origins and of the export statistics of the source countries suggest that fenugreek imports are normally more than 200 tons annually. Only the whole seed is imported and the principal origin is generally Morocco, although recently imports from India have increased substantially. The biggest outlet for fenugreek seed in France is thought to be animal feed flavors with minor uses in spice mixes, retail packs and also for extraction. The usage of fenugreek extracts is mostly in flavor blends but also in some perfumery applications, a little is produced domestically but in addition Indian and Moroccan extract are imported. Moroccan fenugreek extract is produced at the source by a French firm, it is then blended in France, which reexports most of the refined products (Smith, 1982; Commonwealth Secretariat, 1996).

France has one of the highest per capita consumption levels of herbs and spices in Europe. This is due to its high culinary standards, its old colonial ties and its former domestic production base. France is still one of Europe's largest producers of spices and spice extracts.

The market for branded spices is dominated by Ducros (Erdamin Beghin Say) that has more than 50 percent of the market, as well as a major share of the Spanish market. The only major competitor is Amora (Donone group), which has 17 percent of the market and also strong links in Belgium (Liebig Benelux). Supermarket and discount house labels are of increasing importance in France with around 20 percent of the market.

There has been a growing demand for exotic food in France. As a result, sales of specialty spice mixes for Mexican, Thai and Indian cooking have been growing rapidly through specialist companies like Martignon, Laco and Thiercelin (Commonwealth Secretariat, 1996).

The Netherlands

The Netherlands is the third largest spice industry in Europe with a representation of 11.5 percent of the total EU market. The largest immigrant community and the country's old colonies have stimulated the local demand for spices. Fenugreek seed imports are aggregated with turmeric in the Netherlands trade statistics but it is estimated that annual imports are normally around 300 tons. Recently, fenugreek seed imports have increased to 500 tons. The main origin has been Morocco, but recently Moroccan imports have declined and India has become the principal source. Significant quantities of the fenugreek purchased by spice grinders are used in curry mixes. The balance is probably accounted for in animal feed flavors. Several quantities of French-extracted fenugreek absolute are imported annually, mostly for fragrance uses. In addition, smaller quantities of higher strength extracts are produced domestically especially for incorporation into tobacco flavoring, the main markets of which are outside the Netherlands (Spices Board, 1996b; Commonwealth Secretariat, 1996).

The Netherlands is a major re-exporter of spices both to other EU countries and to the USA. It is also a major center for spice processing. Three of the world's largest flavor and fragrance houses have their European manufacturing base there (Quest, International Flavors and Fragrances (IFF) and Tastemaker). All these firms produce oleoresins, essential oils and natural spice extracts using spices imported into the Netherlands. Apart from the above, there are four or five companies specializing in the processing and packing of spices in the Netherlands. These include (owned by Burns and Philip), Conimex (owned by CPC), Van Sillevoldt (Silvo brand) and the Huybregts Groep (Commonwealth Secretariat, 1996).

Prospects for fenugreek seed in its spice application are linked to the demand from domestic curry powder manufacturers. This demand is expected to grow but the increase in terms of volume will be small.

United Kingdom

The UK ranks just behind the Netherlands and Spain as the fifth largest importer of spices in Europe. The UK's historical ties with the Commonwealth, its large Asian and Caribbean ethnic population and its importance in the spice trade ensure its central role in the European spice industry. Fenugreek seed imports are aggregated with those of turmeric in the UK trade statistics, but by means of an examination of origins and of exporting countries statistics it has been estimated that imports have varied between 300–800 tons annually. The peak years were 1976–78, but very recently imports have declined. The principal source has been Morocco, in some years providing 90 percent or more of the total, but lately increasing quantities have been imported from India. China, Israel and Spain have also occasionally supplied smaller amounts. The main use of fenugreek seed is in animal feed flavors. Other outlets for fenugreek seed include curry powders and other spice mixes. There is also some demand for extraction purposes (Smith, 1982; Commonwealth Secretariat, 1996).

The UK is also a major exporter of curry powder, prepared sauces and spicy foods. Food retailing in the UK market is dominated by the supermarkets, which control nearly 70 percent of the market for food stuffs. Most supermarket chains tend to offer only one or two branded spices plus their own label products. Schwartz is the dominant brand with over 50 percent of the market. Three other companies – Lion Food, Bart Spices and British Pepper and Spices (Millstone brand) – together have 16 percent of the market. All the main producers supply their own brand of products for supermarkets, which account for 31 percent of the market (Commonwealth Secretariat, 1996).

The UK is a major center for the manufacture of curry powders, pickles and pre-prepared Asian foods. Companies like Veeraswamy's (part of West Trust), Sharwoods (part of RHM), Trustin Foods and TRS (Sutezwalla) manufacture and export worldwide. The UK is also a major producer of fragrances and flavors with leading multinationals such as Quest International, Bush Boake Alien and specialist firms like Lukas Ingredients and James Dalton (part of the Swiss Flavors house, Firmenich). These companies produce and distribute spice oleoresin, spice oils and a whole range of specialist blend spice extracts and value-added food ingredients. The Seasonings and Spice Association has around twenty-three members, including all the major packets and spice ingredient manufacturers (Commonwealth Secretariat, 1996).

Fenugreek seed remains outside the support system of the European Community's Common Agricultural Policy, there is unlikely to be any inducement for farmers to grow the crop. The UK can therefore be expected to remain a market for imported fenugreek, although no substantial growth is foreseen.

United States of America

No separate import statistics are published for fenugreek seed, but trade sources put imports at about 500 tons annually, with little obvious trend. The main origin is India. Other sources are Morocco, Israel, Pakistan and China. It seems that over half of all imports are used for extraction purposes. Other smaller applications include curry powder and spice mixes. Both solid and liquid fenugreeks are produced domestically by two or three firms. Some Moroccan fenugreek extract is also imported. The extract is mainly used in artificial maple syrups, also in tobacco flavors and some spice seasonings. Demand for fenugreek extract is said to be steady (Smith, 1982; Spices Board, 1996b).

The market is increasingly dominated by two food groups: McCormick Inc. (turnover ECU 1.27 billion) the world's largest spice company and Burns Philip and Co. of Australian (turnover ECU 2.1 billion), which has become through the acquisition of Ostmann in Germany, Euroma in the Netherlands and British Pepper and Spice in the UK, the largest supplier of spices in Europe. These two concerns are estimated to control more that 25 percent of the European market (Commonwealth Secretariat, 1996).

Canada

No separate trade statistics are published for fenugreek seed, but an examination of exporting countries' statistics shows the market size to be about 100 tons annually, imports having remained fairly stable. India is the main country of origin. At one time Morocco was an important supplier but trade informants claim that this source is no longer price-competitive, and very little is now imported from there. The principal uses of fenugreek seed are in spice blends for processed meat products and to a lesser extent, in curry powders. The whole seed is available in retail packs and sales could amount to 10–20 tons annually. Some fenugreek seed may be used for the production of an extract, but Canadian flavor houses generally import a solid extract from USA. It is used entirely in flavors, particularly in artificial maple syrup (Smith, 1982).

Belgium and Luxembourg

Fenugreek seed imports are aggregated with those of turmeric in the Belgium and Luxembourg trade statistics but it is estimated that annual imports varying somewhere between 10–40 tons. The main source is Morocco. The principal uses of fenugreek are in spice mixes and as an animal feed spice. Sales of retail packs of fenugreek seed are minimal, spice packers maintaining that they only stock the line in order to provide a full range of spices. It is unlikely that there will be much growth in the market for a retail pack of the seed (Smith, 1982).

Denmark

Fenugreek imports are aggregated with turmeric in the Danish statistics. However, it is estimated that around 10–15 tons are imported annually, mainly re-exports from Germany, but also small quantities from other European countries and a little directly from Morocco and India. Demand is fairly constant. Mostly whole seed is imported. Nevertheless, importers are willing to take ground fenugreek seed if the quality and price are favorable. The main use for fenugreek in Denmark is in spice mixes but there is also a limited retail trade in whole and ground seed, as well as a small level of usage in medical preparations (Smith, 1982).

Other importers

The other EU member states are not very significant importers of fenugreek and obtain most of their suppliers from other EU states, India and Morocco. Many countries do not publish data for fenugreek imports, but by reference to exporting countries' statistics it appears that the Middle Eastern countries are important markets. *Kuwait* has become a major importer, taking nearly 300 tons. *Saudi Arabia* too, imported over 600 tons annually from India. Other significant importers in the region are *North Yemen* (consistently taking 200–300 tons per annum from India in recent years), the *United Arab Emirates* (averaging about 2,000 tons per annum) and *Oman* (70–80 tons per annum). In North Africa, *Libya* and *Algeria* sometimes take substantial quantities from both India and Morocco. In Asia, Japan is probably the largest market, normally importing between 400 and 800 tons annually from India. Japanese demand for fenugreek seed is mainly for the domestic production of curry powder, and is not expected to show any significant increase. *Singapore's* imports of fenugreek seed have also been around 400–600 tons in recent years, *South Korea* has taken over 200 tons on occasion, while other significant Asian importers include *Sri Lanka*, *Nepal*, *Malaysia*, and *South Korea*. Elsewhere, Australia has occasionally imported over 50 tons per annum, but in most countries fenugreek seed is a very minor spice (Smith, 1982; Spices Board, 1996b).

Trends in consumption and prospects

In the retail markets, spices are generally sold pre-packed in ground or whole form. These usually take the form of glass bottles or cardboard packets. Refills are available for many of the products. In some grocery stores and health food shops spices are sold in open sacks. Customers bring their own containers. More and more spices are being sold in the form of spice mixes or sauces. Pourable sauces is the fastest growing area in the spice retail sector.

There is a continuing debate over the merits and demerits of processing and packing spices at origin. Technically there are few constraints to local processing although tariffs provide some form of trade barrier. The main area of concern is over quality control. Increasingly, stringent food safety laws make it more and more difficult for new producers to afford the cost involved in setting up quality control systems. These have become one of the most important cost elements of the spice trade. The large multinationals like McCormick and Burns Philip encourage processing at source and have set up joint ventures in places like India, Indonesia, Mexico, etc. to provide spices in bulk.

The sale of retail pack spices from the origin is not expected to grow substantially. Apart from the health and safety issue, suppliers need to offer a complete range of perhaps 20–50 different products and must spend very sizeable sums of money on advertising and promotion. This cannot be done from outside the EU and US. An alternative strategy is for producers to invest in processing facilities within these countries.

In the case of spice extracts, particularly oleoresins production at origin is of growing importance. The growth in the industrial processing of spices has paralleled that of the ready-to-eat food and beverage business. Wherever spice flavor ingredients are required for application to food and drink products, spice extracts either in the form of oleoresins, essential oils or occasionally spray dried products are used. The objective of spice processing of this kind is to extract the aromatic and pungent principles from the spices in order to produce a concentrated product of uniform color and flavor. The additional advantage is that product hygiene can be strictly controlled and it can be easily stored and transported.

The spice industry is going through a period of consolidation and concentration. Importing and processing is being handled by fewer and fewer large companies. Many are operating on a European or worldwide scale. The buying power of these companies puts the small grower and exporter at a considerable disadvantage in the bargaining process. To counteract this, growers themselves will have to start working together and build long-term links with these major concerns.

As more and more big European spice houses source their raw materials directly from the countries of origin, there will be increasing contracts between growers and producers and consequently quality controlled growing. Such collaboration can be as joint ventures and involves investment on the part of the spice producer in the country of growth. The advantages for both sides are obvious: increased influence over the raw material quality on the part of the spice processor as well as guaranteed prices, transfer of know-how and technology for the suppliers in the country of origin. Frequently the foreign partner also invests in improved agricultural production facilities and in cleaning and drying and quality control laboratories.

Due to environment and health concerns there has been a growth in the sale of organic spices. There is no doubt that organically certified spices will be seen more and more on the market. At present none of the major brands have entered this field largely because of the lack of assured quality suppliers. Another related development has been that of "diet spices": low sodium, low calories or fat-free sauces and seasonings (Commonwealth Secretariat, 1996; Fotopoulos, 1996).

Elsewhere, the Middle East is fast becoming the major outlet for fenugreek seed and there could be possibilities in the region for new suppliers. The reason for the growth in demand in the Middle East is probably the same as that given for the other spices, namely the influx of migrant workers from South Asia. The other important area where there are prospects for expanded trade is Asia, but imports into many countries in the region varies.

One possible application, for which it is claimed that fenugreek has good prospects, is in the production of diosgenin, a steroid precursor. The main source of diosgenin is wild yams of certain *Dioscorea* sp. Owing to supply problems in the principal producing country, Mexico, diosgenin has become expensive bringing about a switch to cheaper steroid precursors such as steroids from soya beans. This has led to a sharp fall in the proportion of diosgenin to other materials used in the production of steroids. For some oral contraception uses, steroids are also produced by total synthesis nowadays. However, it is most unlikely that the extraction of diosgenin from fenugreek will become economically viable, as a considerable fall in the price of fenugreek would be required, which would also reduce its attraction to growers. Therefore, this usage is thought to offer little prospect for producers.

Acknowledgment

I wish to express my appreciation and thanks to the numerous people who helped to make the completion of this manuscript possible. In particular, Professor Roland Hardman, for his advice and suggestions, the commercial officers of the embassies of India, Spain and Canada in Greece, the directors of the Spices Board, Ministry of Commerce of the Indian Government, the Chamber of Commerce and Industry of Saudi Arabia and the Indian exporting firm, Gautam Export Corporation, for providing me with statistical data and information.

References

Commonwealth Secretariat (1996) *Guidelines for Exporters of Spices to the European Market*, Export Market Development Department, Commonwealth Fund for Technical Co-operation, Commonwealth Secretariat. Marlborough, Pall Mall, London.

Edison, S. (1995) *Research Support to Productivity (Spices)*. The Hindu Survey of Indian Agriculture. pp. 101–5.

Fotopoulos, C. (1995) Total quality management and the Greek food industry. In: K. Mattas, E. Papanagiotou and K. Galanopoulos (eds), *Proceedings of the 44th European Association of Agricultural Economics (EAAE), Seminar on 'Agro-Food Small and Medium Enterprises in a Large Integrated Economy'*, Thessaloniki, pp. 294–4.

Fotopoulos, C. (1996) Strategic planning for expansion of the market for organic products. *Agricultural Mediterranea*, 126, 260–9.

Purseglove, J., Bronon, E., Green, G. and Robbins S.R. (1981) *Spices*, Longman Group Limited, New York.

Robbins, P. (1997) *Tropical Commodities and Their Markets: A Guide and Directory*, Kogain Page, pp. 112.

Smith, A. (1982) *Selected Markets for Turmeric, Coriander Seed, Cumin Seed, Fenugreek Seed and Curry Powder*. Tropical Product Institute publication NG 165, London.

Spices Board (1996a) *What's On*, Ministry of Commerce Government of India, P. B. No. 2277, Cochin.

Spices Board (1996b) *Spices Statistics*, Ministry of Commerce Government of India P. B. No. 2277, Cochin.

Spices Board (1996c) *The Quality People*, Ministry of Commerce Government of India, P. B. No 2277, Cochin.

Spices Board India (1997) *Spice Oils and Oleoresins from India*, Ministry of Commerce, Government of India, P. B. No 2277, Cochin.

Index

T - #0242 - 111024 - C0 - 254/178/11 - PB - 9780367395902 - Gloss Lamination